Lon ...ce
f...
GC ...e

Longman Science for AQA

GCSE Additional Science

Series Editor: **Nigel English**

Muriel Claybrook

Richard Grime

Miles Hudson

Penny Johnson

Sue Kearsey

Colin Lever

Penny Marshall

PEARSON
Longman
Edinburgh Gate
Harlow, Essex

This book also includes

Active Book

Pearson Education
Edinburgh Gate
Harlow
Essex
CM20 2JE
UK
www.longman.co.uk

First published 2007

ISBN 13: 970-1-405-83323-3

Design and production	Roarr Design
Illustration	Oxford Designers & Illustrators Ltd
Picture research	Kay Altwegg
Indexer	Indexing Specialists (UK) Ltd

Printed in China   GCC/01
The publisher's policy is to use paper manufactured from sustainable forests.

Acknowledgments

The publisher would like to thank the following for their help in reviewing this book:

Martin Bendall, Head of Science, Dixons City Academy, Bradford; Stuart Fink BSc (Hons), Head of Science, King Solomon High School, Redbridge; John Kavanagh, Head Of Science, Reigate School; Steve Woolley.

We are grateful to the Department of Transport for use of material from The Highway Code, reproduced under the terms of the Click-Use Licence.

We are grateful to the following for permission to reproduce photographs:

Action Plus: pg166(l) (Neale Haynes), pg168(l) pg179(t) (Glyn Kirk), pg176 pg189(l) (Ingrid Abery); **Agripicture.com:** pg34(r), pg36(t), pg36(bl) (Peter Dean); **Alamy:** pg13(t)(b) (WoodyStock) (m) (Robert Harding Picture Library), pg17(l) (Phototake Inc.), pg22 pg111 (Holt Studios International Ltd.), pg24(t) (Imagebroker), pg26 (Mark Boulton), pg31(b) (Jinny Goodman) (mr) (Paul Glendell), pg42(l) (Otis Images), pg45 (Richard Broadwell), pg51(t) (Jason Pacheco), pg57 (Jeff Morgan), pg58 pg65 pg98 (Phototake Inc.), pg60 (gkphotography), pg81(l) (BananaStock) (r) (Topix), pg94 (Steve Hamblin), pg102(l) (Brand X Pictures) (mr) (Motoring Picture Library) (r) (blickwinkel), pg104 (Coston Stock), pg109 (Nishanm), pg115 (Frank Chmura), pg116 (Topix), pg117 (David Hebden), pg119 (Robert Brook), pg120(t) (Agriculture Images) (b) (David R Frazier Photolibrary, Inc.), pg125 pg210(t) (Leslie Garland Picture Library), pg140 (Rodolfo Arpia), pg142(l) (Martin Anderson), pg144(r) (ScotStock), pg145 (Paul Kern), pg159(l) pg168(r) (Steve Allen Travel Photography), pg160 (Kos Picture Source), pg164 (picturesbyrob), pg166(r) (Image Source), pg167 (Reinhard Dirscherl), pg170 (Blackout Concepts), pg176(t) (Jack Sullivan), pg177 (Stock Image/Pixland), pg190(l) (David R Frazier Photolibrary Inc.) (r) (sciencephotos), pg207(tl) (Ian M. Butterfield), pg212(t) (David Hebden); **Alveyandtowers.com:** pg106, pg209; **Ardea London Ltd:** pg68(r) (John Mason); **Art Directors and Trip Photo Library:** pg97(all) (Helene Rogers); **BAA Aviation Photo Library:** pg193; **Bluegreen:** pg162 (Franck Socha), pg174 (Onne Van Der Wal); **Bubbles:** pg42(r), pg50 (Ian West); **Corbis:** pg48(t) (George Disario), pg73 (Jorge Z. Pasual/epa); **Custom Medical Stock Photo:** pg70(all); **Department for Transport/Paul Brazier and Nigel Worthington at AMV BBDO:** pg172; **Des Kilfeather:** pg189(r); **Empics:** pg61(b) (AP Photo/Jay La Prete), pg128 (Associated Press); **FLPA:** pg28 pg31(tm) (tr) pg68(l) pg150(b) pg151(all) (Nigel Cattlin), pg36(br) (Cyril Ruoso/JH Editorial); **food&drinkphotos.com:** pg46; **Food Features:** pg29(r), pg35, pg56 (Paul David Ellis), pg130, pg144(l); **Garden Picture Library:** pg37 (Howard Rice); **Getty Images:** pg181(r); **Image reproduced with permission from Johnson Matthey:** pg136; **Kimbolton Fireworks:** pg121(l); **Kos Picture Source:** pg217(t) (Carlo Borlenghi); **naturepl.com:** pg31(tl) (Jose B. Ruiz), pg59 (Kim Taylor); **NHPA:** pg32 (David Woodfall), pg34(l) (Martin Wendler),

pg102(ml) (Andy Rouse); **Pearson Education:** pg48(b), pg72, pg87, pg100(b); **Rex Features:** pg208; **Science Photo Library:** pg15(m) (Dr. Torsten Wittmann) (b) (tl) (tr) (Eye of Science), pg17 (Andrew Syred), pg18 (Astrid & Hanns-Frieder Michler), pg23(r) (Photo Insolite Realite) (l) (D. Phillips), pg24(b) (Roger Standen), pg29(l) (Dr. Jeremy Burgess), pg31(ml) (Martyn F Chillmaid), pg36(m) (Veronique Leplat), pg41 (David Hay Jones), pg43 (Steve Gschmeissner), pg44 (Clive Freeman, The Royal Institution), pg49 (Maximilian Stock Ltd.), pg51(b) (Steve Gschmeissner), pg54 (Edward Kinsman), pg61(t) (Sinclair Stammers), pg64(t) (Eye of Science), pg66 (Russell Knightley), pg67 (David Parker), pg71 (Simon Frazer/RVI, Newcastle upon Tyne), pg90 (Charles D Winters), pg94(r) (Astrid and Hanns-Frieder Michler) (b) (Sinclair Stammers), pg100(t) (Pascal Goetcheluck), pg101 (Susumu Nishinaga), pg111(b) (James Holmes, Hays Chemicals), pg113(l) (Ben Johnson) (r) (Charles D Winters), pg121(r) (Andrew Lambert Photography), pg159(r) (Skyscan), pg169(b) (Stephen Dalton), pg175 (Ted Kinsman), pg179(b) (TRL Ltd), pg181(l) (NASA), pg192 (James H Robinson), pg207 (tm) (tr) (bl) (br) pg210(b) (Andrew Lambert Photography), pg212(b) pg215 (Sheila Terry), pg214 (Dr. John Brackenbury); **Sportsbeat Images:** pg53(b); **Still Pictures:** pg194 (Deb Kushal/UNEP); **Sue Kearsey:** pg38; **Warren Photographic:** pg64(b).

The following photographs were taken on commission © **Pearson Education Ltd** by:
Mari Tudor-Jones: pg127, pg137, pg138(all); **Trevor Clifford:** pg20(all), pg81(r), pg86(all), pg98(all), pg103(all), pg105, pg107, pg122(all), pg123(all), pg124, pg133, pg141, pg142(all), pg146, pg147, pg148(all), pg149(all), pg150(t), pg169(t), pg197, pg198, pg201, pg206, pg213(all), pg217(b).

Front cover photos:
Main image: © Oskar Kihlborg/Volvo Ocean Race.
Inset: (top) Mari Tudor-Jones (middle) Punchstock (royalty-free) (bottom) Mari Tudor-Jones.

The publishers are grateful to all the copyright owners whose material appears in this book.

Every effort has been made to trace the copyright holders and we apologise in advance for any unintentional omissions. We would be pleased to insert the appropriate acknowledgement in any subsequent edition of this publication.

How to use this book

This book is divided into three parts, B2 (Biology), C2 (Chemistry) and P2 (Physics). Each of these parts is divided into two units. Each unit has a one-page introduction and is then divided into topics. At the end of each unit there are some practice coursework questions, and at the end of each part there is a set of Foundation and Higher questions that will help you practise for your exams and a glossary of key words.

As well as the paper version of the book there is a CD-ROM called an ActiveBook. For more information on the ActiveBook please see the next two pages.

What to look for on the pages of this book:

Learning objectives
These tell you what you should know after you have studied the topic.

Higher material
If you are hoping to get a grade between A* and C you need to make sure that you understand the bits with this symbol **H** next to it (as well as everything else in the book). Even if you don't think you will get a grade above a C have a go and look at these bits – you might surprise yourself (and your teacher!).

Glossary words
You will need to know the meaning of some key words. These are shown in **bold**. The glossary at the end of each part gives you a list of all the key words and what they mean.

Questions
There are lots of questions to help you think about the main points in each topic.

P2.10

Momentum

By the end of this topic you should be able to:

- calculate momentum
- define momentum
- describe how forces affect the momentum of a body that is moving or able to move
- **H** use an equation to show how force, change in momentum and time taken for the change are related.

An object which is moving has kinetic energy. It also has **momentum (plural: momenta)**. We can calculate the momentum using the equation:

$$\text{momentum} = \text{mass} \times \text{velocity}$$
$$p = m \times v$$
[kilogram metre/second, kgm/s] = [kilogram, kg] × [metre/second, m/s]

Example
What is the momentum of an 800 kg car moving at 12 m/s?

$p = m \times v$
$= 800 \times 12$
$= 9600$ kgm/s

1 Calculate the momentum of:
   a a 742 kg elephant moving at 5 m/s
   b a 70 kg woman skydiving at a terminal velocity of 53 m/s.

2 Would you prefer to try and stop the elephant or the skydiver? Explain your answer.

Momentum is a measure of how much something is moving. It has **magnitude** and direction. This means you must say how much momentum something has, and you must also say what direction the momentum is in.

velocity = 20 m/s   velocity = −20 m/s

A Why have these cars got different momenta?

3 What part of the equation to calculate momentum tells us that we must think about its direction as well as the amount of momentum?

178

6

When a force acts on something that is moving, or able to move, there is a change in its momentum. The force needs to be larger to change momentum more quickly.

H Newton originally worked out that force, change in momentum and time taken for the change in momentum are related:

$$\text{force} = \frac{\text{change in momentum}}{\text{time taken for the change}}$$

$$F = \frac{p_2 - p_1}{t_2 - t_1}$$

$$[\text{newton, N}] = \frac{[\text{kilogram metre/second, kgm/s}]}{[\text{second, s}]}$$

Example

What is the force needed to change a cyclist's momentum by 25 kgm/s in 5 seconds?

$$F = \frac{p_2 - p_1}{t_2 - t_1}$$
$$= \frac{25}{5}$$
$$= 5 \text{ N}$$

B Applying a force will change momentum.

4 Calculate the following:
   a the force needed to change a car's momentum by 3000 kgm/s in 5 seconds
   b the force needed to change a boat's momentum by 400 kgm/s in 8 seconds

5 A 75 kg man starts a sprint race. He increases his velocity from rest to 3 m/s in 1 second. What force must he be using to do this?

Car safety

C Why do you fly forwards if your car stops suddenly?

Seatbelts help to save people's lives in an accident. If you are not strapped in, your momentum keeps you moving if the car stops suddenly. You will continue to move forwards inside the car and could hurt yourself by hitting the dashboard or windscreen. The seatbelt keeps you in position. Losing your momentum very quickly needs a large force and this can injure you. In a sudden stop, the seatbelt stretches a little, reducing your momentum over a longer time, so the force on you is lower.

An airbag in a car can reduce your momentum to zero, so you stop moving. By doing this slowly, the force on you is less than if you hit the hard dashboard and stopped suddenly.

6 Write a short story about the Animalympics. Include details of the masses and velocities of five different animals in a race. Include at least one of the animals changing velocity and explain how this affects the momentum and how it might have happened. At the end, explain how to calculate the momenta of each animal and then put them in order of increasing momentum.

H Add a calculation of the force needed to cause the momentum change in your story.

How science works
Some sections of a topic might focus on science in everyday life or the practical skills involved in science. These are highlighted where they appear in each topic.

Summary exercise
The answer to the last question in the topic summarizes the whole of the topic. The exercise will also be useful for revision.

How to use your ActiveBook

The ActiveBook is an electronic copy of the book, which you can use on a compatible computer. The CD-ROM will only play while the disc is in the computer. The ActiveBook has these features:

DigiList
Click on this tab and all the electronic files on the ActiveBook will be listed in menus.

ActiveBook tab
Click this tab to access the electronic copy of the book.

Glossary
Click this tab to see all of the key words and what they mean. You can read them or you can click 'play' and listen to someone else read them out for you to help with the pronunciation.

Key words
Click on any of the words in **bold** to see a box with the word and what it means. You can read it or you can click 'play' and listen to someone else read it out for you to help with the pronunciation.

Interactive view
Click this button to see all the bits on the page that link to electronic files. You have access to all of the features that are useful for you to use at home on your own. If you don't want to see these links you can return to **Book view**.

ActiveBook | DigiList | bc Glossary

B2.23

Different kinds of cells

By the end of this topic you should be able to:

- describe what cell differentiation is and when it happens in animals and plants
- explain what stem cells are
- give examples of how stem cells might be used to treat diseases
- list some of the social and ethical issues about using embryonic stem cells.

A There are many kinds of cell in your body that do different jobs.

Your body contains many different kinds of cells, and almost all of them are **differentiated** with special shapes to do different jobs. When an animal egg cell is fertilised and starts dividing to make an **embryo**, the cells in the early stages can differentiate into any kind of cell. However, once cells are differentiated, they can only make more of the same kind of cell when they divide. These new cells can only be used to replace old cells or repair damage.

1 a Write a list of as many different kinds of cell in your body as you can think of.
  b Add to your list the job that each kind of cell does.

2 a Why can a muscle cell only divide to make more muscle cells?
  b Why do differentiated body cells divide to make more of the same kind of cell?

3 A cell in an early embryo can differentiate into any kind of cell. Explain why.

60

 © Pearson Education 2007 | Turn off | Go Interactive | Page

Target sheets
Click on this tab to see a target sheet for each unit. Save the target sheet on your computer and you can fill it in on screen. At the end of the topic you can go back and see how much you have learnt by updating the sheet.

Help
Click on this tab at any time to search for help on how to use the ActiveBook.

get sheets   ?   Help

In plants, it is different. Many cells in many parts of a fully grown plant keep the ability to differentiate into any kind of cell.

4 Explain why we can grow whole new plants from leaf, root or shoot cuttings, but not a complete human from an arm or leg.

Undifferentiated animal cells are called **stem cells**. The cells in an early embryo are called 'embryonic stem cells'. There are also stem cells in tissues in your body, such as in your **bone marrow**. These are difficult to extract but, when they have been extracted, they can be treated to turn into many kinds of differentiated cells.

5 a Write down two sources of stem cells.
  b Suggest which would be the easier source to get stem cells from.
  c Explain your answer.

B In tissue culture, we take undifferentiated cells from the tip of a plant shoot, and treat them so that they develop into differentiated cells in leaves, shoots and roots.

Stem cells for treatment

People suffer from many diseases that are caused by cells not working properly or being damaged. There is a lot of research taking place into how embryonic stem cells could be used to make replacement cells for ones that are not working properly. For example, nerve cells grown from stem cells could cure **paralysis**, and insulin-secreting cells grown from stem cells could one day cure type 1 diabetes. However, some people argue that embryos have the potential to become whole beings and so should not be used like this.

C Tyson Gentry (an American footballer) was paralysed when an accident damaged nerve cells in his spine. Now he cannot walk.

6 a How could stem cells be useful in the future?
  b Suggest why some people argue against embryonic stem cell research.

7 A quote in a newspaper about a paralysed person says: 'Stem cells could make me walk again'. Explain how this could be done and why stem cells make it possible.

61

Zoom feature
Just click on a section of the page and it will magnify so that you can read it easily on screen. This also means that you can look closely at photos and diagrams.

61 of 256

Page number
You can turn one page at a time, or you can type in the number of the page you want and go straight to that page.

Contents

Growing our food

As the world's human population increases, so it makes increasing demands on food production. Growers in countries such as the UK have to find ways of producing a large range of foods at competitive prices. Often this means **factory farming** to give high yields of fruit, vegetables, meat and other food products. This way of growing can change the land and the air around us.

There are other environmentally friendly ways of producing food, such as organic and free-range farming, that do less damage to the environment. However, can they provide enough food for everyone at a price we are prepared to pay?

Consumers also expect to have most fresh fruits and vegetables, such as strawberries and green beans, available all year round. This means either importing them from countries thousands of miles away or growing them in greenhouses when our weather is unsuitable. We could just eat food when it is in season, but are we prepared to have a much more limited choice of fresh food?

A Is this the only way to produce enough food?

By the end of this unit you should be able to:

- describe the structure and function of the parts of animal and plant cells
- explain how dissolved substances get into and out of cells by diffusion and osmosis
- explain how plants obtain the food they need to live and grow
- explain what happens to energy and biomass within a food chain
- explain that waste material produced by plants and animals decays and describe the conditions needed for decay to happen
- outline the carbon cycle.

1 What do you think is meant by:
   a factory farming
   b organic farming?

2 List all the benefits of intensive farming of chickens that you can think of for:
   a the farmer
   b the customer.

Animal cells

By the end of this topic you should be able to:

- give the structure and function of the parts of animal cells
- describe what controls chemical reactions in cells
- give examples of how animal cells may be specialised to carry out a particular function.

Your body is made up of millions of cells. Each cell is surrounded by a very thin **cell membrane** which holds the cell together. The cell membrane also controls what goes into and out of the cell.

Each cell contains smaller parts called **organelles**. These include the nucleus, mitochondria and ribosomes. The organelles have particular jobs in the cell. The **nucleus** controls the cell's activities. It is surrounded by **cytoplasm**. Many chemical reactions take place in the cytoplasm. These reactions are controlled by chemicals called **enzymes**.

Mitochondria are organelles that use glucose in **respiration** to release energy for the cell. **Ribosomes** are the smallest organelles. They build up or synthesise proteins. Proteins are molecules that are used to make other parts of the cell and other chemicals such as enzymes. The cells of most animals have the same organelles.

1 Write a definition of the word 'organelle'. Give examples.

2 What controls which substances enter and leave the cell?

3 Which organelles use glucose to release energy?

4 Ribosomes synthesise proteins. Explain what this means.

5 Explain why the nucleus of the cell is important.

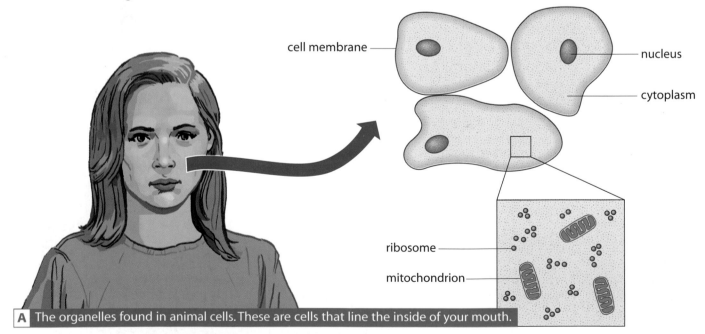

cell membrane

nucleus

cytoplasm

ribosome

mitochondrion

A The organelles found in animal cells. These are cells that line the inside of your mouth.

The cells in your body are not all the same. They have different shapes and many have special features that are related to what they do. These cells are called **specialised cells**. Some examples are shown in diagram B.

a Muscle cells have fibrils and can shorten in length.

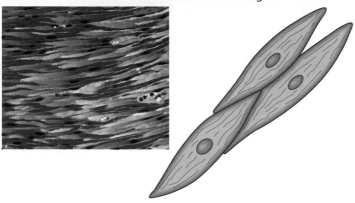

b Sperm cells have a tail to help them move to find the egg. They also have a high number of mitochondria (energy).

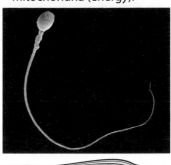

c Nerve cells have long fibres that carry electrical impulses.

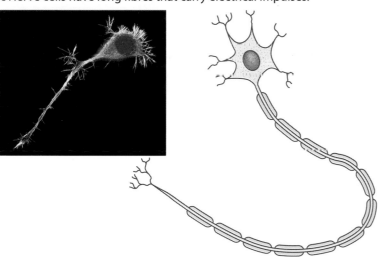

d Epithelial cells cover surfaces inside and outside the body. These epithelial cells have cilia that move constantly back and forth to move particles along.

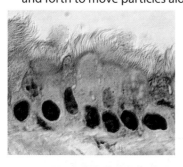

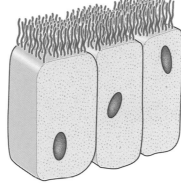

B Examples of specialised cells.

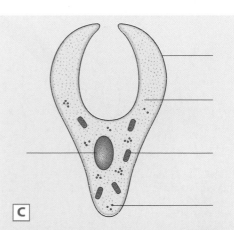

C

6 Look at the photographs in B.
- **a** Give two differences between a muscle cell and a ciliated epithelial cell.
- **b** Most cells don't move. Explain why a sperm cell needs a tail.
- **c** Explain how the features of a nerve cell are adapted for its job.
- **d** The epithelial cells shown help to sweep mucus containing dust out of the lungs. Describe how they are adapted for this job.

7 The cell shown in diagram C is called a 'goblet cell'. It is found in the lining of the stomach where it releases mucus.
- **a** Suggest one way in which it is specialised for the job it does.
- **b** Copy the diagram of the cell. Complete the labels to show the different parts of the cell and what they do.

P
D
P
P

Plant cells

By the end of this topic you should be able to:

- describe the structure and function of the parts of plant cells
- give examples of how plant cells may be specialised to carry out a particular function.

Plant cells also usually have a cell membrane, nucleus, mitochondria, ribosomes and cytoplasm. Unlike animal cells, plant cells also have a **cell wall**. The cell wall gives a plant cell strength and support.

Some plant cells also have organelles called **chloroplasts** in their cytoplasm. Inside the chloroplasts is a green pigment called **chlorophyll**. Chlorophyll is a chemical that plants use in **photosynthesis** to absorb the Sun's light energy. Photosynthesis produces glucose.

In the centre of many plant cells there is a large permanent space called a **vacuole**. The vacuole is filled with a liquid called **cell sap**, which contains sugars, salts and water.

1 Why is the cell wall important in a plant cell?

2 Explain why plants need:
  a a nucleus
  b mitochondria
  c a cell membrane
  d ribosomes.

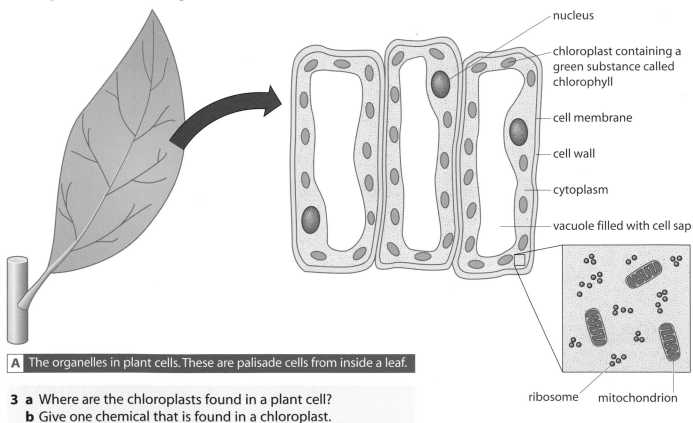

nucleus

chloroplast containing a green substance called chlorophyll

cell membrane

cell wall

cytoplasm

vacuole filled with cell sap

ribosome   mitochondrion

A The organelles in plant cells. These are palisade cells from inside a leaf.

3 **a** Where are the chloroplasts found in a plant cell?
  b Give one chemical that is found in a chloroplast.
  c Why are chloroplasts important to plants?

4 **a** Where is the cell sap found in plant cells?
  b What does cell sap contain?

Many plant cells are specialised to carry out particular jobs.

- **Palisade cells** are found in the leaf and are packed with chloroplasts.
- **Root hair cells** have extensions into the soil to absorb water and dissolved mineral ions.
- **Xylem vessels** are made up of dead cells, arranged as long tubes with no end walls between them. They transport water from the roots, up through the stem to the leaves.

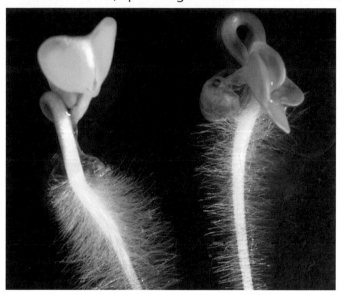

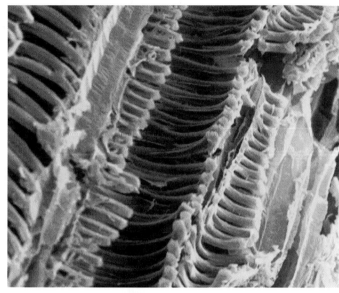

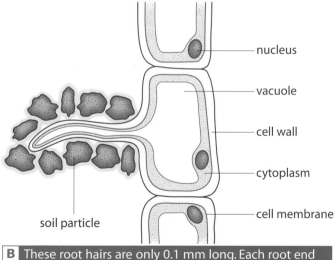

- nucleus
- vacuole
- cell wall
- cytoplasm
- cell membrane
- soil particle

B These root hairs are only 0.1 mm long. Each root end has thousands of these cells.

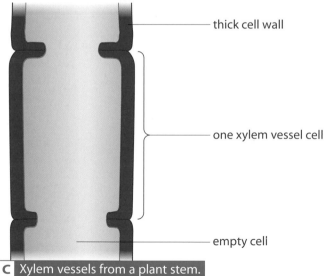

- thick cell wall
- one xylem vessel cell
- empty cell

C Xylem vessels from a plant stem.

5 Why do palisade cells have lots of chloroplasts?

6 Suggest which parts of a plant might not contain chloroplasts. Explain your answer.

7 Explain how the root hairs of cells are adapted for their job.

8 a What is the only cell structure present in the xylem cell?
  b How does this lack of other organelles help it to transport water up a stem?

9 Look at diagram A on page 14 and diagram A on page 16.
  a Make a list to show how animal and plant cells are similar.
  b Write down any differences between animal and plant cells.

Cells make tissues make organs

By the end of this topic you should be able to:

- relate the structure of different types of cells to their function in a tissue or an organ.

Some living things are made up of only one cell. Other living things are made from millions of cells. In more complex organisms, specialised cells of the same type group together to form **tissues**. Cells in a tissue might make a thin sheet, like the epithelial cells that make up linings inside the body. Cells in other tissues group together, like muscle cells that make muscle tissue.

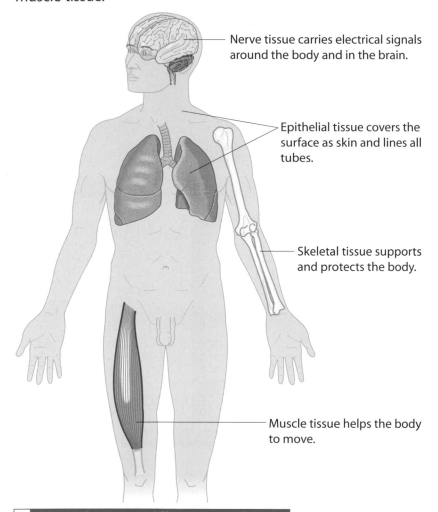

Nerve tissue carries electrical signals around the body and in the brain.

Epithelial tissue covers the surface as skin and lines all tubes.

Skeletal tissue supports and protects the body.

Muscle tissue helps the body to move.

B Humans have many different types of tissues.

Groups of tissues join together to make more complicated structures that are called **organs**. For example, your arteries are organs. They are lined with epithelial tissue and surrounded by muscle tissue and fibrous tissue. Other organs in your body include your heart, brain, liver, stomach and lungs.

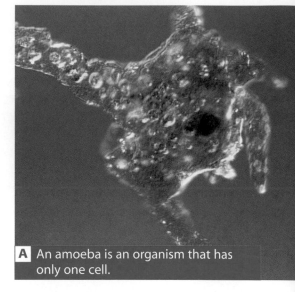

A An amoeba is an organism that has only one cell.

1 Write down the organelles you would expect to find in an amoeba.

2 What is the job of:
   a bone tissue
   b nervous tissue?

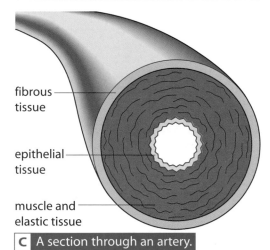

fibrous tissue

epithelial tissue

muscle and elastic tissue

C A section through an artery.

3 Write down six organs in your body.

4 Suggest what the jobs of the following tissues are in an artery:
   a epithelial tissue
   b muscular tissue.
   (*Hint*: look back at Topic B2.1 for help.)

5 Why do we describe an artery as an organ?

Specialised plant cells also group together to make tissues, such as epidermal tissue, photosynthetic tissue and vascular tissue. The organs of a plant include the root, stem and leaves.

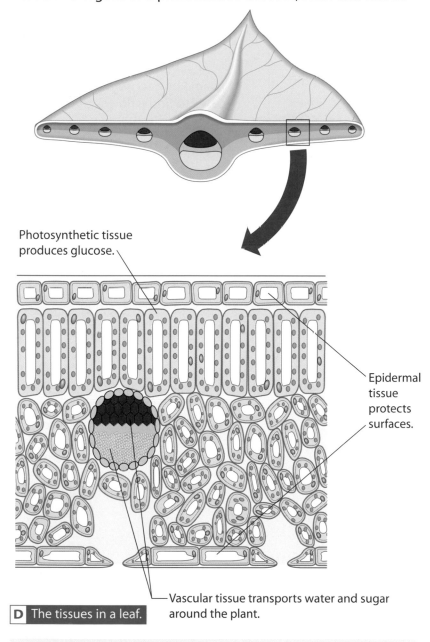

Photosynthetic tissue produces glucose.

Epidermal tissue protects surfaces.

Vascular tissue transports water and sugar around the plant.

D The tissues in a leaf.

6 Suggest one kind of plant cell that will be found in vascular tissue. Give a reason for your answer.

7 Complete table E for plants.

Tissue type	Function	Special features of cells in tissue
Photosynthetic		
	transport of water	
	protection of surface	

E

8 The stomach is an organ in your body. Its job is to churn food around to mix it with chemicals from the stomach wall that help digest the food. Copy diagram F, which shows a section through your stomach. Label it to show:
  a at least two types of tissue and the cells they contain
  b the job of the different tissues and how the cells in them are adapted for that job.

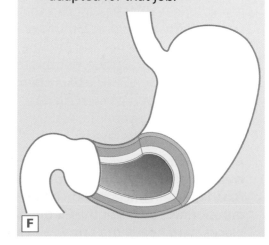

F

Diffusion

By the end of this topic you should be able to:

- describe the process of diffusion
- give an example of diffusion through cell membranes
- explain how dissolved substances can move into and out of cells
- describe how the difference in concentration can affect the rate of diffusion
- explain what a partially permeable membrane is.

1 Name one sense that a butterfly uses to find a flower that is far away.

2 Explain how the perfume of the flower reaches the butterfly.

3 Where is the concentration of perfume particles highest: near the flower or far from it? Explain your answer.

You can smell the perfume released by a flower because smelly particles spread through the air. The smell can spread many metres. The movement of the smelly perfume particles through the particles of air is called **diffusion**.

As you get closer to the flower the smell gets stronger. This is because there are more smelly particles near the flower. We say that the **concentration** of perfume particles is higher nearer the flower.

A This butterfly is guided to the flower by the scent.

Diffusion can also happen in liquids. When a **soluble** substance is placed in water, the particles that make up the substance will start to diffuse. The particles move in random directions, and bump into each other and into the water particles. They start all clumped in one place, but this movement spreads them out slowly. When the particles are clumped together, they have a high concentration. When they are more spread out, they have a lower concentration. So as they spread more, the concentration gets lower.

A difference in concentrations of a substance between two areas is called a **concentration gradient**. If you start with a much greater concentration in one place than the other, then diffusion will be faster than if the concentrations in the two places were nearly the same.

12 noon

6 pm

6 am

B Colour diffuses from a slice of boiled beetroot in water.

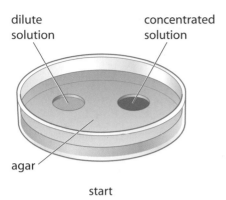

dilute solution

concentrated solution

agar

start

one day later

More coloured particles have diffused into the agar from the concentrated solution than from the dilute solution. The concentrated solution has a faster rate of diffusion.

C The greater the concentration gradient, the faster diffusion takes place.

All cells have a cell membrane. The cell membrane has tiny holes through which small particles can pass by diffusion. We say that cell membranes are **partially permeable membranes** because large particles can't get through.

If the concentration of small particles on each side of a membrane is different, then more particles will diffuse through the membrane from the concentrated solution to the dilute solution.

partially permeable membrane

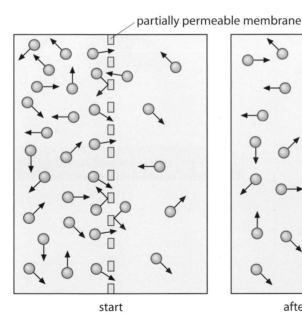

start

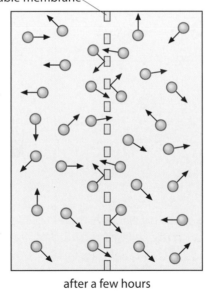

after a few hours

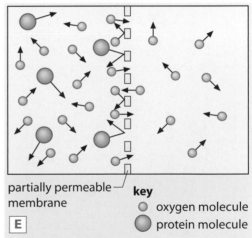

partially permeable membrane

E

key
- oxygen molecule
- protein molecule

D When the concentrations are different, the **net movement** of particles is from the higher concentration to the lower concentration.

4 Explain what we mean by 'partially permeable membrane'.

5 Explain how cell membranes control the particles that pass through them by diffusion.

6 Explain what we mean by 'net movement'.

7 Look at diagram E. Oxygen molecules are small molecules, but protein molecules can be very large.
  a Cells need oxygen for respiration. Explain how oxygen gets into cells. Use the words 'diffusion' and 'partially permeable membrane' in your explanation.
  b When cells use oxygen in respiration, the concentration of oxygen gets lower in the cell. Explain why this means more oxygen will diffuse into the cell.

Osmosis

By the end of this topic you should be able to:

- explain that osmosis is the movement of water across a partially permeable membrane from a dilute solution to a more concentrated solution
- describe how water moves between cells by osmosis.

1 What happens to a plant if it isn't watered?

Water molecules are small enough to diffuse through partially permeable membranes. A dilute solution contains more water molecules than a concentrated solution (see diagram B). If a partially permeable membrane separates two solutions of different concentrations, there will be net movement of water molecules from the dilute solution to the concentrated solution. Diffusion of water molecules across a partially permeable membrane is called **osmosis**.

A A plant without water wilts. If a plant is kept watered it stands upright.

partially permeable membrane

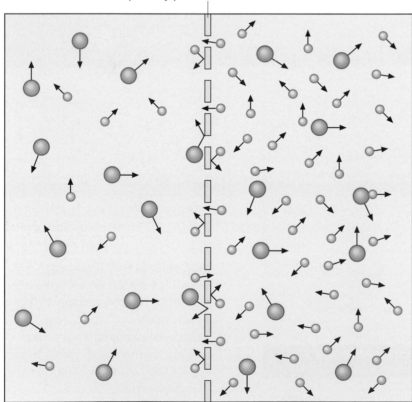

concentrated solution   dilute solution

key

⚬ water molecule   ⬤ solute molecule

B Osmosis occurs across a partially permeable membrane.

2 Write a definition for the word 'osmosis'.

3 Explain why net movement of water molecules in diagram B is from the dilute solution to the concentrated solution across a partially permeable membrane.

Osmosis explains the movement of water molecules across plant and animal cell membranes. In a plant, water moves from the soil into root-hair cells, because soil water is a more dilute solution than the solution in the cytoplasm of the cell. The extra water molecules in the root hair cell dilute its concentration compared with cells inside the root. The water molecules then diffuse into those cells, which make their concentration more dilute. In this way water moves from cell to cell into the centre of the root by osmosis.

4 Look at the photographs in A.
  a Describe how water is moving through the roots of the well watered plant.
  b If you compared the concentration of the cytoplasm of a root cell from the wilted plant with the concentration of the cytoplasm of a root cell from the watered plant, which would have the higher concentration? Explain your answer.

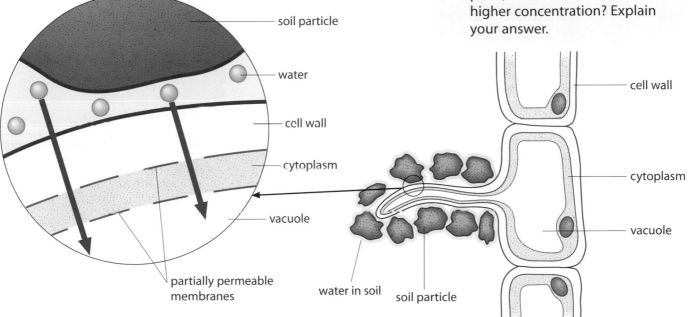

soil particle

water

cell wall

cytoplasm

vacuole

partially permeable membranes

water in soil

soil particle

cell wall

cytoplasm

vacuole

➡ osmosis from soil into root hair

C Water moves into a plant by osmosis.

Osmosis also happens between animal cells. However animal cells have no strong cell wall to protect them, so if too much water enters them they burst. If too much water leaves them they shrink and cannot function properly.

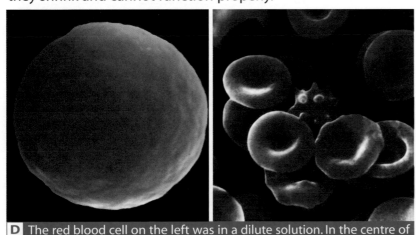

D The red blood cell on the left was in a dilute solution. In the centre of the right-hand photo ia a spiky red blood cell that was in a concentrated solution.

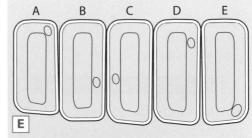

A   B   C   D   E

E

5 In the plant cells above the concentration of sugar and salts changes from A to E. A is the least concentrated and E is the most concentrated.
  a Copy diagram E.
  b Add an arrow underneath the cells to show the direction water will travel from cell to cell by osmosis.
  c Colour the partially permeable membranes in red.
  d Explain why water moves from cell to cell.

Photosynthesis

By the end of this topic you should be able to:

- summarise photosynthesis in an equation
- describe what happens during photosynthesis
- list factors that may limit photosynthesis.

'Synthesis' means to combine or to join together. Photosynthesis uses light energy to join together two simple molecules, carbon dioxide and water, to make a more complex molecule called glucose. Oxygen is also released as a **by-product**.

A These plants are photosynthesising.

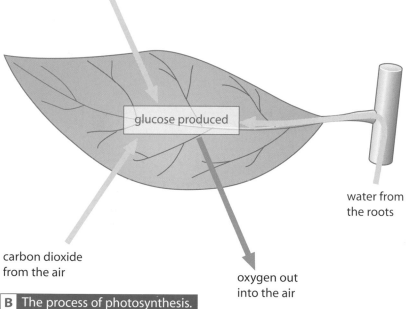

light energy from the Sun

glucose produced

water from the roots

carbon dioxide from the air

oxygen out into the air

B The process of photosynthesis.

Photosynthesis happens in a series of reactions that can be summarised by the equation:

carbon dioxide + water (+ light energy) → glucose + oxygen

1 a What are the reactants in photosynthesis?
b Where do they come from?
c Why is light needed for photosynthesis?

2 a Explain what we mean by a 'by-product'.
b Give an example of a by-product.

3 A farmer forgets to water the crops when the weather is dry. What effect will this have on photosynthesis?

Chlorophyll is a green pigment that is found in chloroplasts. The chlorophyll absorbs light energy which converts carbon dioxide and water into glucose.

4 Chloroplasts are found mainly in palisade cells in the upper layer of leaves. Suggest a reason for this.

C

5 Photograph C shows a plant with variegated leaves.
a Suggest what is missing in the white areas of the leaves.
b Explain what this means for photosynthesis in these leaves.

The speed at which photosynthesis takes place is the **photosynthesis rate**. It is affected by the environment. The factors which affect it most are:

- temperature
- availability of carbon dioxide
- light intensity.

If the level of one or more of these is low, the rate of photosynthesis will be slowed down or **limited**. The factor that is reducing the rate of photosynthesis is called the **limiting factor**. Increasing the amount of a limiting factor will increase the rate of photosynthesis. However, increasing temperature increases the rate of photosynthesis only up to a point when the temperature causes the enzymes that control the reactions to break down.

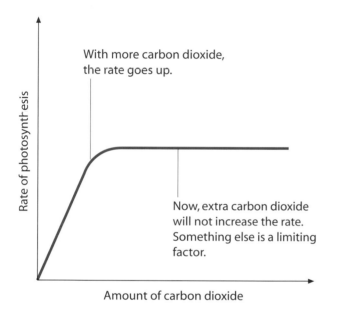

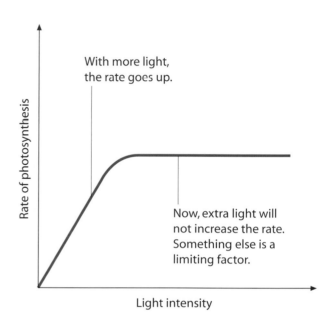

D The amount of carbon dioxide and the light intensity affect the rate of photosynthesis.

6 Look at the graphs in D.
  a Suggest why increasing the amount of carbon dioxide increases the rate of photosynthesis.
  b Suggest why increasing the amount of light can increase the rate of photosynthesis.
  c Explain why both graphs level off as the factor continues to increase.
  d The percentage of carbon dioxide in the air is about 0.04%. On a warm sunny day, suggest which factor is limiting the rate of photosynthesis in the middle of a crop. Give a reason for your answer.

7 Pondweed is a plant that lives under water.
  a Write a word equation for the process that pondweed uses to make food.
  b Suggest three ways that you could use in the school lab to make pondweed grow faster.
  c Some tiny plants grow on the surface of ponds. If a pond gets completely covered by these the larger plants underneath them die. Suggest why.

Improving crop yields

By the end of this topic you should be able to:

- recall that factors which limit the rate of photosynthesis can interact with one another
- evaluate the benefits of changing environments to improve growth rate
- interpret data on the limiting factors of photosynthesis.

A Plant growers grow crops in polytunnels and greenhouses so they can control some of the factors affecting the rate of photosynthesis.

The rate of plant growth depends on the rate of photosynthesis. The rate of photosynthesis is controlled by factors such as amount of carbon dioxide, temperature and light intensity. These factors can interact with each other to limit the rate of photosynthesis.

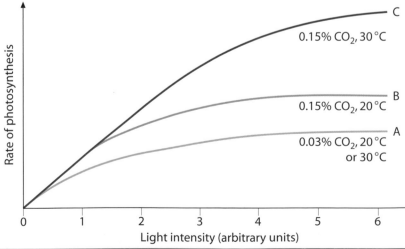

B How light intensity, temperature and level of carbon dioxide interact to affect rate of photosynthesis.

1 Give three factors that can be controlled in polytunnels and greenhouses to increase the rate of photosynthesis.

2 Look at graph B.
   a Which curve shows the highest rate of photosynthesis?
   b Which curve shows the lowest concentration of carbon dioxide?
   c Explain why curve C is much higher than curve B.
   d Suggest why you get curve A even if the temperature is increased from 20 °C to 30 °C.)

Growing crops in greenhouses and polytunnels makes it possible to control the environment around the plants. Sheltering the plants from a cool wind or heating the air round the plants increases the temperature. In greenhouses, light intensity can be increased on dull days by using artificial lighting. Carbon dioxide levels in a greenhouse can be increased by burning a fuel or adding the gas.

The faster the rate of photosynthesis, the better the growth of the plants. The better the growth of plants the higher the **yield**. Yield is the amount of a crop that a plant produces.

3 For each of the factors in your answer to question 1, explain how a grower can increase that factor in a greenhouse.

4 Why would a grower want to increase these factors?

5 Suggest one other factor needed for photosynthesis that a grower could control more easily in a greenhouse.

C Homegrown tomatoes in winter.

Artificially controlling the environment around a crop also makes it possible to grow a crop at times of year that it wouldn't grow as well, or in places that it wouldn't grow well. The disadvantage of growing crops in this way is the cost of the energy for providing heating, lighting and additional carbon dioxide. However, it results in a bigger yield, and crops grown out of season usually sell for more money.

6 a Give three advantages of growing crops in controlled environments.
  b Give one disadvantage of growing crops in controlled environments.
  c Suggest why a grower might choose to grow crops in a controlled environment.

7 UK-grown lettuces are available at Christmas time. Explain how this is possible. Include:
  a what conditions the lettuces need to grow well
  b how these conditions are provided in winter
  c what the conditions are like outdoors at this time
  d the advantages and disadvantages to the lettuce grower of supplying lettuces in December.

Energy stores in plants

By the end of this topic you should be able to:

- explain that the glucose produced by photosynthesis can be converted to starch and stored
- explain what plant cells use glucose for.

During photosynthesis plants make glucose:

carbon dioxide + water (+ light energy) ⟶ glucose + oxygen

The glucose a plant makes does not stay as glucose for long. Some of it is used for respiration:

glucose + oxygen ⟶ carbon dioxide + water (+ energy)

A Potato plants use some glucose in respiration for growth and some to make potatoes.

1 Look at the equations for photosynthesis and respiration.
 a What is the same in the two equations?
 b What is different in the two equations?
 c Explain the difference between the two equations.
 d Animals also respire. Where do they get their glucose from for respiration?

The energy released in respiration in plant and animal cells is used in many processes. It is used to build new cells, for new growth or for repair of damaged cells. Some energy is used in chemical reactions to change some materials into others. More energy is used to move materials around inside the organism.

A plant won't use all the glucose it makes straight away. Some glucose needs to be converted to starch and stored for times when the plant can't make enough.

2 Give three processes for which a plant needs the energy from the respiration of glucose.

3 Suggest two times when a plant can't make all the glucose it needs. (*Hint*: look back at the photosynthesis equation.)

Starch is useful for storage because it is **insoluble** and doesn't react easily with other chemicals in the cytoplasm. The starch is an **energy store**, because the plant can change it back to glucose when it needs more energy from respiration.

cell wall

cytoplasm

starch grain

B Starch is stored in plant cells as starch grains in the cytoplasm.

Most plant cells contain some starch, but some cells are specially adapted to store a lot of it. For example, potatoes are stem tubers full of starch which grow underground. In the spring, these stores provide the energy needed to make leaves for new potato plants. Seeds, such as those of wheat and rice plants, also store a lot of starch. These stores are also used to provide the energy needed to make new leaves when the seed germinates.

C Many of our energy foods come from parts of plants that are stores of starch.

4 Explain what we mean by energy store.

5 Name:
  a one plant where starch is stored in stem tubers
  b two plants where starch is stored in the seeds.

6 Explain why potato plants and plants growing from seed need food stores.

7 Suggest why plants that store starch are grown as crops by farmers.

8 A wheat grain contains a lot of starch.
  a Explain where the starch came from.
  b Explain why the starch is stored in the grain.
  c Write down what the starch will be used for by the plant.

Minerals and plant growth

By the end of this topic you should be able to:

- describe how plants get the mineral ions they need for healthy growth
- explain what plants use nitrates and magnesium for
- describe the symptoms of plants deficient in nitrate or magnesium ions.

Carbon, oxygen and hydrogen make up about 96% of a plant, but a plant also needs some **mineral ions** to grow well.

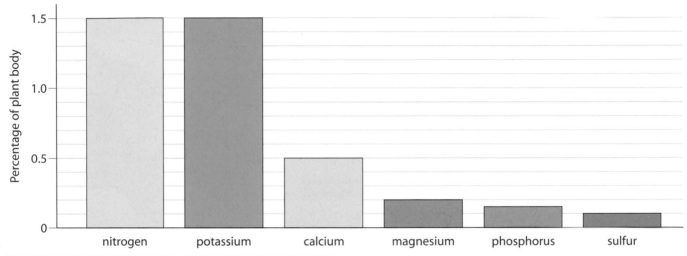

A Percentages of mineral ions found in an average plant.

The mineral ions are dissolved in soil water. Plants absorb the mineral ions through their root hair cells. Many mineral ions, like magnesium, are simple ions of the element. However, nitrogen does not form ions easily, so plants get the nitrogen they need from nitrate ions in the soil.

Plants need the different mineral ions for different processes. Both nitrogen and magnesium are needed to make chlorophyll. If the plant can't get enough of a mineral ion, it will not grow well. If a plant is lacking a mineral, it will show particular **deficiency symptoms**.

1 Look at graph A. Which two mineral ions are most common in plants?

2 How much magnesium do plants usually contain?

Mineral ion	Use	Deficiency symptoms
Nitrate	Used to make amino acids which are used to build proteins. Important element in chlorophyll production.	Leaves turn yellow and growth is stunted.
Magnesium	Important element in chlorophyll production.	Leaves turn yellow.

B Deficiency symptoms.

C This plant is healthy. | **D** This plant is deficient in nitrate. | **E** This plant is deficient in magnesium.

Fertilisers

As a plant grows it absorbs mineral ions from the soil. If the amount of mineral ions in the soil gets low, then more mineral ions need to be added. This may also be needed after the plants have been harvested if the farmer wants to plant a new crop in the same place.

Mineral ions can be replaced by adding **fertilisers** to the soil. Fertilisers are chemical mixtures that contain the mineral ions that plants need for healthy growth. A farmer might use natural fertilisers like animal manure or compost, or artificial fertilisers like ammonium nitrate.

F Different kinds of fertilisers can be added to soil to help crops grow.

3 a Name two ions that are needed to produce chlorophyll.
  b Name one ion that is needed to make amino acids.

4 Explain why a lack of nitrate or magnesium ions will reduce plant growth.

5 How would a farmer tell if a plant was not getting enough:
  a nitrate   **b** magnesium?

6 Explain why the amount of mineral ions in the soil decreases when a crop is harvested.

7 How can a farmer increase the amount of mineral ions in the soil?

8 A farmer notices that the leaves of the cabbages in a field are turning yellow. The plants are also smaller than expected.
  a Which two mineral ions could the plant need more of? Explain your answer.
  b Describe where these ions come from and how they get into the plant.
  c What could the farmer do to make his cabbages healthier? Explain your answer.

31

Biomass and energy flow

By the end of this topic you should be able to:

- describe the role of plants in capturing the Sun's energy and how this energy is stored
- explain how biomass changes at each stage in a food chain
- interpret and construct pyramids of biomass
- describe how energy is lost at each stage in a food chain and the reasons for this.

During photosynthesis, a plant captures light energy using chlorophyll. The energy is used to produce molecules of glucose and is released during respiration. Some glucose is changed to other substances that make up the plant, and some into stores of starch until the glucose is needed for respiration again. This increases the **biomass** (mass of living material) of the plant.

A **food chain** shows what eats what. It also shows how energy passes from one organism to another. When the cow eats grass, energy stored in the substances in the cells of the grass plants is transferred to the cow. If we eat meat from the cow, some of the energy stored in the substances in the cells of the cow is transferred to us.

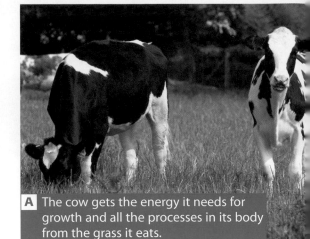

A The cow gets the energy it needs for growth and all the processes in its body from the grass it eats.

1 Explain how photosynthesis increases:
   a the energy stored in a plant
   b the biomass of a plant.

2 How does a cow increase its biomass?

grass (producer)　　　　　cow (primary consumer)　　　　　human (secondary consumer)

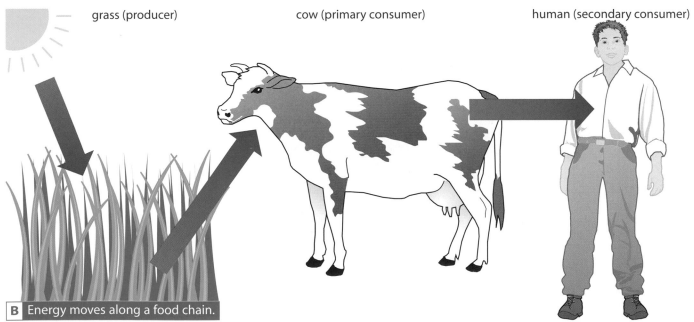

B Energy moves along a food chain.

3 Suggest what we mean by:
   a producer   b consumer.

4 Look at diagram B. Why is the human a secondary consumer here?

5 Explain why the Sun is the source of all the energy in this food chain.

At each stage in the food chain, not all of the energy that the organism takes in is changed into biomass (see diagram C). Some of it is lost again as heat to the environment during respiration. In animals, some of the food is never digested and is lost as faeces from the body. And some of the substances from the body are lost in urine. An animal also loses energy as heat when it moves about.

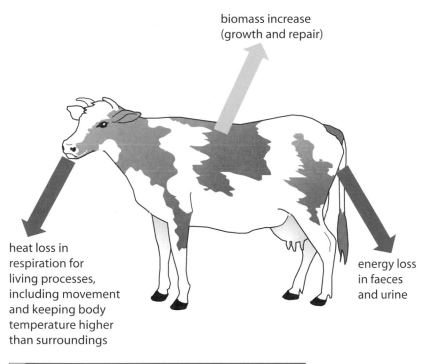

biomass increase
(growth and repair)

heat loss in respiration for living processes, including movement and keeping body temperature higher than surroundings

energy loss in faeces and urine

C Biomass increases and energy losses from a cow.

A **pyramid of biomass** shows the relative amount of living material at each level in a food chain. The bottom level is always the producer, and each level up shows the next level in the food chain.

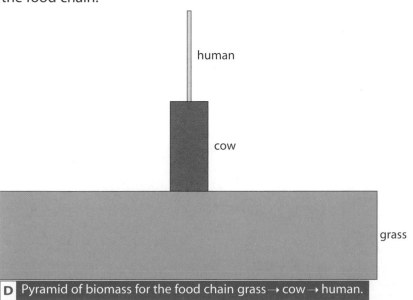

human

cow

grass

D Pyramid of biomass for the food chain grass → cow → human.

6 Give three reasons why the biomass of the cow is less than the biomass of all the plants it eats.

7 Sketch a diagram to show what happens to the energy in your food.

8 Explain why a pyramid of biomass gets narrower as you go up each level.

9 Owls eat mice, and mice eat seeds.
  a Draw a food chain to show this relationship.
  b Explain where the energy comes from at the start of the food chain.
  c Draw a pyramid of biomass for this food chain.
  d Explain the shape of your pyramid.

Producing more food

By the end of this topic you should be able to:

- describe how the efficiency of food production can be improved
- evaluate the effects of managing food production and distribution.

The human population continues to increase, so we need to produce more food to feed everyone. We could do this by changing what we eat. Eating at a lower level in the food chain means there is more available to eat because the higher up the chain you eat at, the more energy has already been lost to the environment.

1 a At what level are we in the food chain if we eat cows?
  b At what level are we in the food chain if we eat plant crops?
  c Use the pyramid to explain why we would have more food if we ate plant crops.

2 Suggest one reason why we don't just eat plant crops.

Farmers can also increase the amount of energy that their animals use to make biomass by reducing some of the ways that energy is lost. This makes the food production more **efficient**.

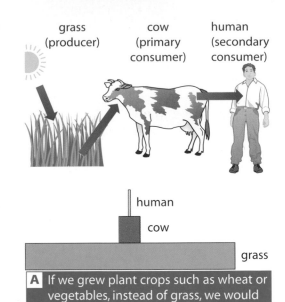

grass (producer)   cow (primary consumer)   human (secondary consumer)

human
cow
grass

A If we grew plant crops such as wheat or vegetables, instead of grass, we would have more food to eat.

B These hens are kept in artificial light and can hardly move, so they produce more eggs.

C Some poultry are raised in large sheds where heaters keep the shed warm.

If animals are kept in controlled surroundings, they can also be fed special food mixtures containing the right balance of nutrients. This can help them grow faster.

There are disadvantages to producing food more efficiently. Crowding animals together increases stress, and can make it more difficult to keep the animals healthy, because it is easier for disease to spread. Chemicals, such as antibiotics, are added to their food to reduce the chances of them becoming ill. It costs money for fuel for the heaters in heated sheds, and this increases the cost of producing the food. Some people think that it is cruel to keep animals in these conditions.

from caged birds

D There are different costs to producing each kind of egg.

A farmer also has to consider other costs, such as how much it will cost to transport produce to where it will be sold. It will be cheaper to transport produce to a local farmers' market than to send it to a large supermarket far away. But the supermarket may be able to sell more of the produce more quickly.

6 A farmer wants to decide whether to keep battery hens or free-range hens.

  a Copy and complete table E to help the farmer decide.

	Advantages	Disadvantages
Battery hens		
Free-range hens		

E

  b Suggest other factors the farmer should think about before making a decision.

3 Look at photographs B and C.
  a Explain why reducing animal movement makes food production more efficient.
  b Give another method that farmers can use to make food production in animals more efficient.
  c Birds and mammals lose a lot of heat to their surroundings because they keep their internal temperature higher than the surroundings. Explain the advantage of keeping them warm.

4 How can controlling the food given to animals help a farmer?

5 Look at photograph D. Choose two kinds of eggs and give one advantage and one disadvantage of producing eggs that way.

Death and decay

By the end of this topic you should be able to:

- explain how materials are returned to the environment
- explain why materials decay
- list the conditions that speed up decay
- describe how the decay process releases substances that help plants to grow
- explain how, in a stable community, the same material is constantly recycled.

Living organisms are made up of large complex molecules such as proteins and carbohydrates. Plants build these complex molecules using materials they get from the air and soil. Animals make the materials they need from the food that they eat.

When an organism dies, or an animal gets rid of waste materials such as urine or faeces, the complex molecules don't just stay as they are. Sometimes the dead bodies are eaten by **scavengers**. Small particles of dead plant and animal material may be eaten by **detritus feeders**. **Microorganisms** start to grow on the rest of the dead organisms and waste materials. These microorganisms are called **decomposers** and include bacteria and fungi.

A Why is the world not covered in dead bodies?

1 Draw a flow diagram that shows how a nitrate molecule in the soil becomes a protein molecule in a plant that is passed along a food chain.

B Mould is a kind of fungus.

C Earthworms feed on **detritus**.

D Crows are scavengers.

2 a Name three types of organism that feed on dead plant and animal material.
  b Give one example of each type.

Microorganisms are very important because they digest large complex molecules, such as proteins. They break them back down into smaller molecules. This releases the mineral ions that formed them, such as nitrate ions, back into the soil. These are the ions that new plants need for healthy growth.

3 If there were no microorganisms, what would happen when an organism died?

4 Explain why decomposers are important in keeping soil fertile for healthy plant growth.

5 We describe the movement of materials through plants, animals and microorganisms as a cycle. Explain why.

When a **community** of plants, animals and microorganisms is stable, the materials that are removed from the environment are replaced by those returned by decomposers. The materials are constantly recycled.

The rate of growth of microorganisms is also affected by the conditions they are growing in. They usually grow fastest in warm, damp conditions. Many kinds of microorganisms also grow faster when there is plenty of oxygen for respiration.

E Compost heaps provide ideal conditions to decay dead plant material.

6 Our food is dead plant and animal material. Explain why putting it in a fridge helps to keep it for longer.

7 Suggest the best conditions for making compost as quickly as possible.

8 Using the diagram you started for question 1, add microorganisms to show what happens when plants and animals in the food chain die. Your diagram should show how the protein molecules are broken down to nitrate ions again, and what happens to the nitrate ions.

The carbon cycle

By the end of this topic you should be able to:

- describe how carbon dioxide is removed from and returned to the environment
- explain how the carbon is made into complex molecules and passes along the food chain
- describe how all the energy originally captured by green plants has been transferred when all the materials in the cycle have been decomposed to plant nutrients again.

Think about a molecule of carbon dioxide in the air. It enters a plant leaf and is combined with water during photosynthesis to make glucose. The glucose may be used in respiration, which releases the carbon back to the air in a carbon dioxide molecule. Or the glucose may be changed into other molecules that make up the plant.

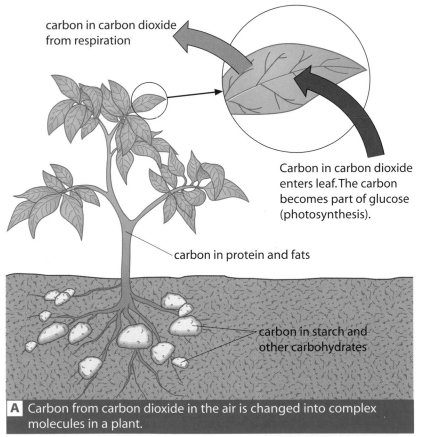

carbon in carbon dioxide from respiration

Carbon in carbon dioxide enters leaf. The carbon becomes part of glucose (photosynthesis).

carbon in protein and fats

carbon in starch and other carbohydrates

A Carbon from carbon dioxide in the air is changed into complex molecules in a plant.

Now think about what happens if the plant is then eaten by an animal. In the animal, the plant material is digested, absorbed into the animal's body and then used to make complex molecules such as fats and proteins. Some of the plant material also supplies glucose which the animal uses for respiration. Any plant material that isn't digested is excreted as faeces by the animal.

1 a Give three kinds of complex molecule in plants that contain carbon.
  b Give one simple molecule in the environment that contains carbon.

2 Explain where the carbon in a starch molecule in a potato originally came from.

B Faeces, such as this cow pat, contain many partially digested complex molecules.

3 Give two kinds of complex molecules in animals that contain carbon.

4 Some carbon is released from animals as a simple molecule.
  a What is this molecule?
  b How is it produced?

5 Explain why faeces are good food for detritus feeders and decomposers.

6 Describe what happens to the carbon in complex molecules in an animal when it is eaten by another animal.

Microorganisms and detritus feeders break down the dead bodies and waste products of organisms. They use their food to build more body tissue, but they also respire. When they respire, some of the carbon in the food they eat returns to the environment as carbon dioxide. Eventually the carbon dioxide will enter another plant and be converted into glucose again. This cycling of carbon through organisms is called the **carbon cycle**.

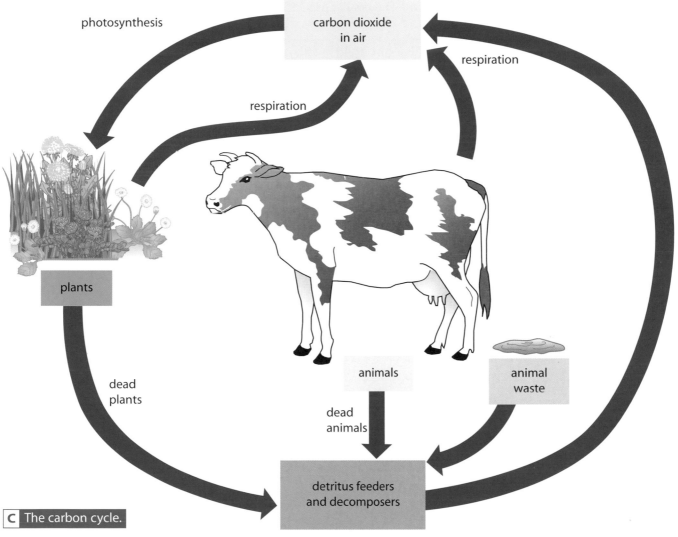

C The carbon cycle.

At the same time as the carbon cycle takes place, the light energy that was captured from the Sun by chlorophyll and used to make glucose and other complex molecules is also moving along the food chain. Unlike carbon, the energy is not cycled. At each stage in the food chain, energy is lost as heat to the environment. By the time that microorganisms have recycled the carbon and plant nutrients back to the environment, all the energy that the plants originally captured has been transferred to the environment as heat.

7 Explain why energy is lost to the environment at each stage in a food chain.

8 a Draw a sketch of diagram C.
b Identify the molecules that carbon is found in at each stage.
c Add arrows to your diagram to show how energy moves through this system.

Investigative Skills Assessment 1

These questions refer to an experiment carried out by a student to investigate how the rate of osmosis is affected by concentration.

The students used two different concentrations of sucrose (20% and 40%). They measured the rate at which the solution in the capillary tube rose over a 15 minute period. The results are shown in the table below.

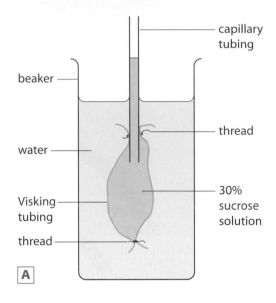

A

Time (minutes)	Height of 20% sucrose solution (cm)	Height of 40% sucrose solution (cm)
0	0.0	0.0
1	1.0	2.0
2	1.5	3.8
3	2.0	5.0
4	2.5	6.5
5	3.0	8.0
6	3.0	9.2
7	3.4	10.4
8	3.6	11.4
9	3.7	12.6
10	3.8	13.6
11	4.0	14.4
12	4.0	15.4
13	4.0	16.0
14	4.0	16.6
15	4.0	17.0

B

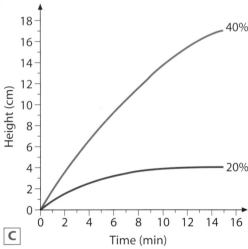

C

1 Name the **two** variables that are being measured. *(2 marks)*

2 What is:
a the dependent variable *(1 mark)*
b the independent variable? *(1 mark)*

3 a Name **two** other variables that will have to be kept the same to make the test fair. *(2 marks)*
b For each variable described in part **a**, say how it could affect your results. *(2 marks)*

4 Looking at the data from the more dilute sucrose solution, which is the anomalous result? *(1 mark)*

5 From the data what appears to be the relationship between concentration and osmosis? *(2 marks)*

6 What would you do to make the investigation more reliable? *(2 marks)*

7 ✎ Explain in your own words what happens to the height of the sucrose in each experiment over time. *(2 marks)*

8 Write a suitable conclusion for this investigation to explain the relationship between concentrations and:
a rate of diffusion *(1 mark)*
b the time taken to reach the point when net osmosis is zero. *(1 mark)*

Total = 17 marks

Treating diabetes

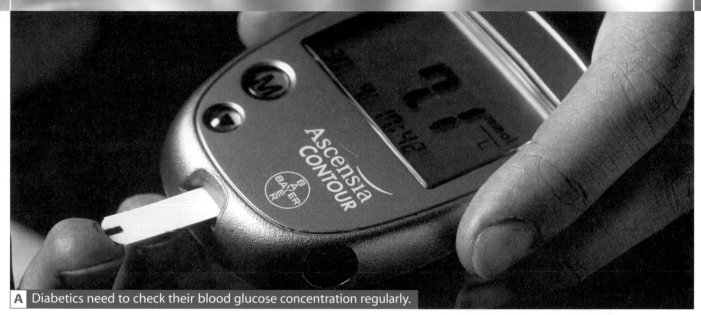

A Diabetics need to check their blood glucose concentration regularly.

When we digest our food, enzymes break down the carbohydrates to simple sugars like glucose. These are absorbed into the blood and carried around the body to where they are needed for respiration and other processes. Other enzymes inside our cells are used in many of these processes.

After a meal the concentration of glucose in the blood is high. If it remained like this, it would damage the body. Normally a hormone called insulin is released into the blood. This causes the blood glucose concentration to return to normal. However some people have a disorder called diabetes where the blood glucose concentration is not controlled properly.

There are over 1.6 million people in the UK with this disorder and the number is rising. It has many causes: some genetic, some environmental. The kind of treatment needed depends on the type of diabetes and how severe it is. In the future, stem-cell treatment and embryo screening could help treat or prevent this disorder.

1 Read these statements and sort them into the following groups: 'I agree', 'I disagree', 'I want to find out more'.
  - Enzymes inside the body catalyse reactions such as respiration and digestion.
  - Enzymes that are extracted from organisms are very useful.
  - Our bodies have to control internal conditions otherwise we could die.
  - Genetic disorders can be inherited.
  - Stem cells and embryo screening should be used in medical research and treatment.

By the end of this unit you should be able to:

- define what enzymes are and describe some of their functions
- describe how your body keeps some internal conditions constant
- give examples of human characteristics that show a simple pattern of inheritance
- make informed judgements about the use of stem cells and embryo screening.

Essential respiration

By the end of this topic you should be able to:

- describe how aerobic respiration in mitochondria uses glucose and oxygen to release energy
- explain what cells use this energy for
- summarise aerobic respiration with a word equation.

A All these activities need aerobic respiration.

Aerobic respiration literally means respiration using air. More accurately, it uses oxygen from the air. The reactions of respiration are going on in your cells all the time. Without these reactions you would die. They provide all the energy that your body uses to stay alive and to do all that you do.

1 Make a list of ten things that you use energy for. Try to include some that allow you to stay alive.

2 a How does the oxygen needed for respiration get into your body?
  b Glucose is a kind of sugar. Where does the glucose for respiration come from?

The reactions of aerobic respiration can be summarised using this word equation:

glucose + oxygen → carbon dioxide + water + energy

All your cells need energy, so they all respire aerobically. Although aerobic respiration can be summarised by the equation on page 42, there are many reactions in respiration. Most of these reactions happen inside small structures called **mitochondria**.

Mitochondria are inside all cells, and they all need oxygen and glucose to do their job. Both oxygen and glucose are carried round the body in the blood.

3 a Describe how oxygen gets from outside the body to the inside of every cell.
   b Describe how glucose gets from the food you eat to the inside of every cell.
   c Respiration produces carbon dioxide and water. Suggest what happens to these products to stop them building up in the body.

4 A muscle cell has more mitochondria than a brain cell. Suggest a reason for this.

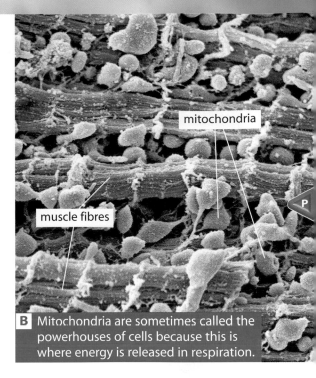

B Mitochondria are sometimes called the powerhouses of cells because this is where energy is released in respiration.

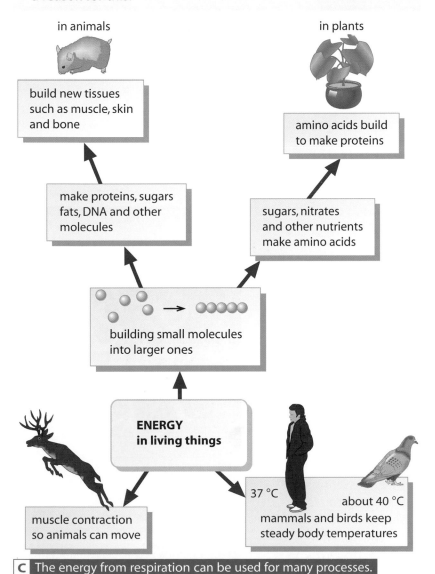

in animals

build new tissues such as muscle, skin and bone

make proteins, sugars fats, DNA and other molecules

in plants

amino acids build to make proteins

sugars, nitrates and other nutrients make amino acids

building small molecules into larger ones

ENERGY in living things

37 °C
mammals and birds keep steady body temperatures
about 40 °C

muscle contraction so animals can move

C The energy from respiration can be used for many processes.

5 a What do both plants and animals use energy for?
   b What else do animals need energy for?

6 Pat is breathing faster than Sanjay. Give as many reasons as you can to explain this.

7 Copy the equation for aerobic respiration. Add notes to show:
   a where the reactants come from
   b where the products go to in your body
   c what the energy is used for.

43

Speeding up reactions

By the end of this topic you should be able to:

- explain that enzymes are biological catalysts that speed up (catalyse) reactions
- give some examples of processes that enzymes catalyse in living cells
- describe how enzyme structure relates to the way it works.

A Chemical reactions make the substances your body needs to grow bigger.

During a chemical reaction the substances that are reacting are rearranged to make new substances. We can speed up the rate of some reactions by using a **catalyst**. The catalyst makes it easier for the reacting substances to come together and be rearranged, so the reaction happens faster.

Catalysts speed up many of the reactions that happen in living organisms. These biological catalysts are called **enzymes**.

1 What is a catalyst?

2 Describe what would happen if we did not have enzymes in our bodies.

Enzymes are proteins. Proteins are large molecules made of a long chain of **amino acids**. Each kind of enzyme has a different order of amino acids in the chain. The chains of amino acids fold up in a way that is determined by the order of the amino acids.

Enzymes are very special because their shape allows the substances of a reaction to fit into the enzyme. This brings the substances closer together and in the right position to react quickly. When they have reacted, they no longer fit the space on the enzyme, and so they leave. This makes the space free for more reacting substances to fit into the enzyme. The enzyme is not changed by the reaction.

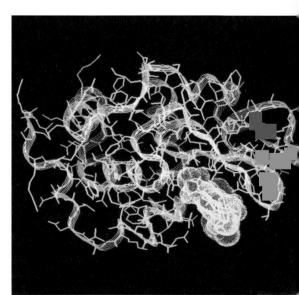

B This model of an enzyme shows the amino acid chain (blue sticks) and the shape that they fold into (pink ribbon). One of the reacting substances (yellow) fits neatly into a special gap in the enzyme shape.

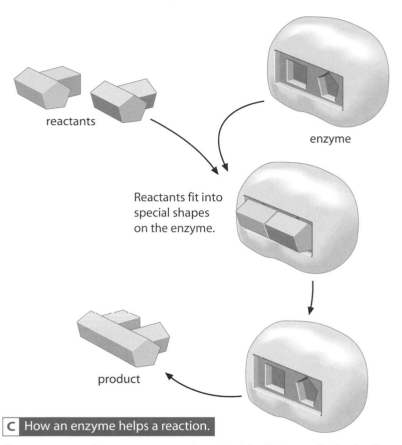

reactants

enzyme

Reactants fit into special shapes on the enzyme.

product

C How an enzyme helps a reaction.

Enzymes catalyse many reactions inside living cells, including those of respiration. They speed up the reactions that build large molecules such as proteins from their building blocks of amino acids. In plants they also catalyse the reactions in photosynthesis. Each different reaction needs a different enzyme because each reaction involves different substances.

Different enzymes work best in different conditions, such as at different temperatures and at different pH (acid or alkaline). The body needs to control temperature and pH so that the enzymes work at their best all the time.

D These yellow bacteria are growing around very hot, acidic water springs.

3 The shape of an enzyme affects the way it works. Explain why.

4 High temperatures destroy the shape an amino acid chain makes.
  a What effect would high temperatures have on an enzyme-controlled reaction?
  b Explain your answer.

5 Explain why every reaction needs a different enzyme.

6 Bacteria contain enzymes that help speed up their reactions. Suggest the conditions that the enzymes in hot-spring bacteria work best at.

7 a Make a list of all the reactions that you know are catalysed by enzymes.
  b Describe how a plant would be affected if it contained no enzymes.

Enzymes in digestion

By the end of this topic you should be able to:

- name enzymes that work outside body cells in digestion and where they are released from
- give examples of the reactions in the digestive system
- describe the conditions that different digestive enzymes work best in.

A Your food contains proteins, starches and sugars, fats and oils.

The molecules of proteins, starch and fats are made of thousands of small units. They are huge – much too large for you to absorb into your body as they are. This means they have to be **digested** (broken into small units) in your gut. These digestion reactions need to be quick so that you can absorb what you need from your small intestine, before the remains pass out of your body. All of these reactions are catalysed by enzymes to speed them up.

Different enzymes work on different substances:
- **amylase** catalyses the breakdown of starch to sugars
- some **protease** enzymes catalyse the breakdown of protein into smaller molecules, and others catalyse the breakdown of those molecules to amino acids
- **lipase** enzymes catalyse the breakdown of **lipids** (fats and oils) into **fatty acids** and **glycerol**.

These enzymes are made in specialised cells in **glands** in different parts of your gut. They are made inside cells, but they move out of the cells into the gut where they work. Some mix with the food in the gut, others remain attached to the outside of cells in the gut wall.

1 There are many kinds of enzyme in your gut. Explain why.

2 What would happen if you didn't have enzymes in your gut?

3 Look at diagram B.
   a In which two parts of the gut is amylase made?
   b Protease enzymes are made in the small intestine wall. Where else are they made?
   c In which part of the gut does lipid digestion take place?

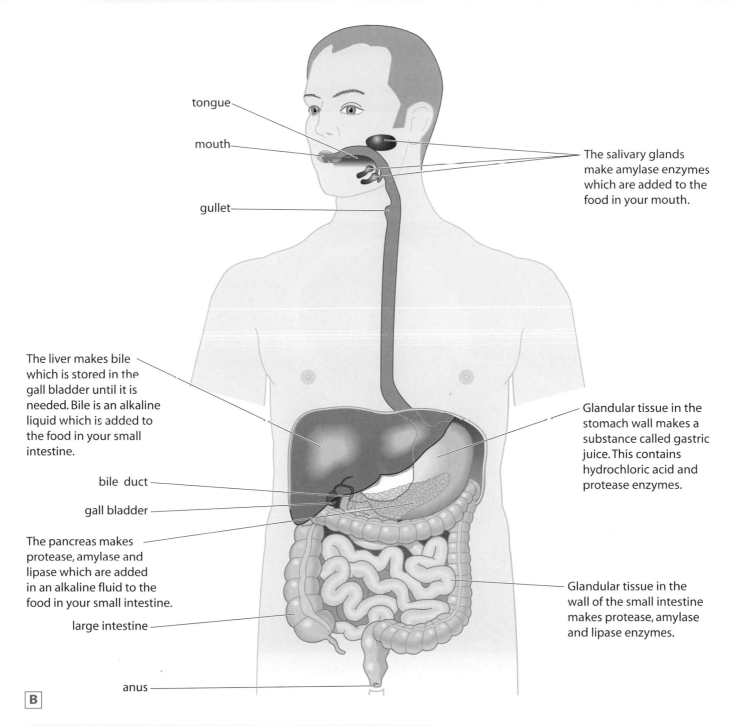

tongue

mouth

The salivary glands make amylase enzymes which are added to the food in your mouth.

gullet

The liver makes bile which is stored in the gall bladder until it is needed. Bile is an alkaline liquid which is added to the food in your small intestine.

Glandular tissue in the stomach wall makes a substance called gastric juice. This contains hydrochloric acid and protease enzymes.

bile duct

gall bladder

The pancreas makes protease, amylase and lipase which are added in an alkaline fluid to the food in your small intestine.

Glandular tissue in the wall of the small intestine makes protease, amylase and lipase enzymes.

large intestine

anus

B

4 Look at diagram B. The stomach makes a chemical that is not an enzyme.
  a What is the chemical?
  b Suggest the conditions that enzymes in the stomach work best in.

5 a Describe how bile reaches the small intestine.
  b What is one role of bile in digestion?
  c Explain why bile is needed to do this.
  d Bile also acts as a detergent, making the fat drops much smaller. Suggest how this helps the enzymes to work better.

6 You eat a cheese sandwich. Write down the stages of digestion of the fat and protein in the cheese and the starch in the bread as they pass through your gut. Include which enzymes and other chemicals are needed in each part of the gut for successful digestion.

Useful enzymes

P

By the end of this topic you should be able to:

- explain why enzymes are used in biological detergents
- describe how microorganisms produce enzymes that pass out of their cells and state some industrial uses of these enzymes
- evaluate the use of enzymes in the home and in industry.

A How will these clothes ever be clean again?

The dirt that we get on our clothes comes from our bodies, our surroundings and from the food we eat. The substances that make up the dirt are mostly proteins, fats and sugars. Before we understood about enzymes, clothes were usually washed at high temperatures to help remove stains containing fats. These high temperatures would make stains containing proteins stick to the material. So the protein stains had to be treated with chemicals first to break down the protein.

P

1 Suggest the problems of washing before enzymes were used in washing powders. Consider:
   a the stages needed to get clothes clean
   b the energy needed to heat the water to get clothes clean
   c the time needed to get the clothes clean.

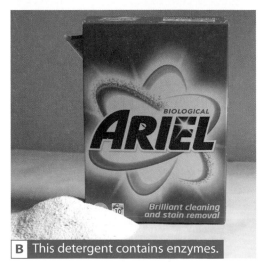

B This detergent contains enzymes.

Evaluating enzyme use

Biological washing **detergents** now contain many enzymes, such as proteases, lipases and amylases. These enzymes work best at a temperature of about 40 °C. Some people still use non-biological detergents because they get an **allergic reaction**, such as a skin rash, to the enzymes.

The enzymes that we use for washing clothes come from bacteria that have been grown in large vats. The enzymes pass out of the bacterial cells and are collected from the liquid in the vat.

In some baby foods, proteases are used to help 'pre-digest' the proteins. This makes the food easier for the baby to digest.

It is cheaper to get starch from plants than sugars to use in our food. Starch can be converted to sugar syrup using **carbohydrases** to catalyse the reaction. This means we can make products like this sports drink more cheaply.

Glucose makes foods sweeter but we can convert this to fat if we eat more than we use in respiration. Glucose can be converted to **fructose** using **isomerase**. Fructose is much sweeter than glucose, so less is needed to sweeten foods for slimming.

D Examples of other enzymes we collect from bacteria for use in industry.

4 a Give three examples of enzymes used in the food industry.
 b For each of your examples, give one advantage of using it.

2 What do the following enzymes digest:
 a protease
 b lipase
 c amylase?

3 Look at your answers to question **1a–c**. Suggest the advantages of using enzymes to help get clothes clean.

C Millions of bacteria can be grown quickly in this vat. Enzymes extracted from the liquid are used for many purposes.

5 Great-aunt Ethel washed her clothes using biological detergent on a very hot wash at 90 °C. They still came out with stains on. Explain to her:
 a why her clothes would have been washed cleaner if the water temperature had been 40 °C
 b the advantages of using enzymes in washing detergents
 c one disadvantage of using enzymes in washing detergents.

Clearing out the waste

By the end of this topic you should be able to:

- name some of the waste products that must be removed from the body
- explain how they are removed from the body.

A As well as breathing out water vapour, which we can see on a cold day, we also breathe out carbon dioxide.

The reactions in your body produce substances that the body needs, but they also produce substances that your body has no use for. Some of these **waste products** could actually damage cells or interfere with other reactions if too much collects in your body. So your body has to get rid of, or **excrete**, these waste products.

1 a Which process in your body produces carbon dioxide as a waste product?
  b Where in your body is the carbon dioxide produced?
  c Where in your body is the carbon dioxide excreted?

Carbon dioxide dissolves in water to make an acid. Your body contains a lot of water, and many of the reactions that happen in your body take place in this water. Enzymes control many of these reactions, and enzymes are affected by pH (acidity). This means that the amount of carbon dioxide in your body must be controlled so that the enzymes can work properly.

Another substance your body excretes is **urea**. Urea is made from the breakdown of amino acids that the body doesn't need. The body can't store excess amino acids so they have to be removed. They travel in the blood to the liver where they are broken down and used to make urea. Urea has to be removed because it is **toxic** and could harm the body.

2 What would happen to the pH of your body if carbon dioxide from respiration was not excreted?

3 Suggest what effect this would have on enzyme-controlled reactions in your body. (*Hint*: look back at Topic B2.15.)

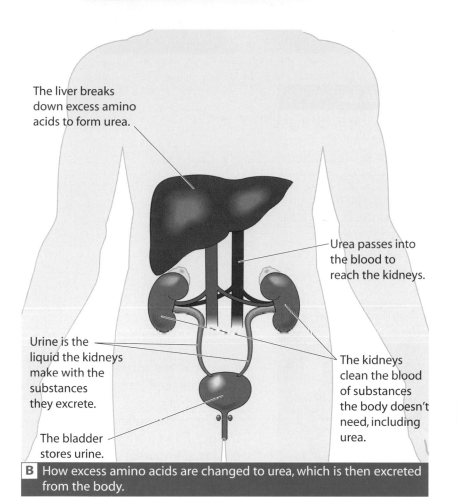

The liver breaks down excess amino acids to form urea.

Urea passes into the blood to reach the kidneys.

Urine is the liquid the kidneys make with the substances they excrete.

The kidneys clean the blood of substances the body doesn't need, including urea.

The bladder stores urine.

B How excess amino acids are changed to urea, which is then excreted from the body.

C This dog is emptying his bladder of urine.

4 a Amino acids are building blocks for bigger molecules. What are these bigger molecules?
  b Where does the body get the amino acids it needs to make these bigger molecules?
  c What does the body use these bigger molecules for?

5 Look at diagram B and photo C. Draw a flow chart to show how excess amino acids are removed from your body.

6 Look at photograph D.
  a Name one substance that is being removed from the patient's body by the dialysis machine.
  b Name one substance that the patient is excreting without using the dialysis machine.
  c Explain why these substances must be removed from the body.

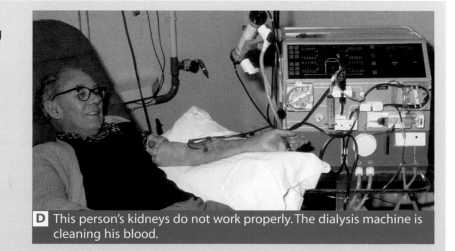

D This person's kidneys do not work properly. The dialysis machine is cleaning his blood.

P
D
P
P

Keeping a balance

By the end of this topic you should be able to:

- give examples of some internal conditions in the body that are controlled
- explain how and why water and ion content are controlled.

In Topic B2.18 you learnt how concentrations of carbon dioxide and amino acids are controlled in the body to protect cells and the reactions inside them from damage. The body needs to control the concentrations of other substances too, if the cells and their reactions are to work properly. Keeping everything in balance in the body is called **homeostasis**.

A Eating and drinking add substances to your body.

1 a List all the substances that you have taken into your body in the last 24 hours.
b List all the substances you have got rid of from your body in the last 24 hours.
c Try to identify any substances in list **a** that are not included in list **b**.

2 Write a definition for homeostasis.

Food and drink don't just contain the sugars, proteins and fats that get broken down and absorbed into your body. They also provide water and **mineral ions**, such as salt (sodium chloride). These are all small molecules and easily pass across cell membranes. Too much or too little of these in your body will cause problems.

B Too much water in your body is called **overhydration**. Too little water in your body is known as **dehydration**.

Too much water in your cells can cause them to swell and even burst. Too little water can slow down reactions, and make it difficult for other substances to move around. The movement of water in and out of cells is also linked to the ion content in the cells. If the ion content is not at the right concentration, then the water content won't be either because too much may move into or out of the cells.

3 Look at photograph C.
   a Describe the difference between the two cells.
   b Explain the difference between the two cells,
   c Suggest what would happen if the cell on the left was put into pure water. Explain your answer.

Your body controls the balance of water and ions in your blood by removing excess water and ions through the kidneys.

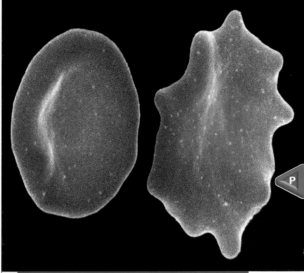

C Two red blood cells: the one on the left is normal, the one on the right has been in a solution that contained a lot of mineral ions and has lost water.

D Water is also lost from the body through sweating.

4 For each of the following, say if the person could become dehydrated or overhydrated. Give a reason for each answer.
   a Jenny doesn't have a drink all day.
   b Denny drinks two litres of water at once.
   c Benny eats a large pack of salted crisps.
   d Penny sits outside for 2 hours on a hot summer day.

5 a Describe two ways that water is lost from the body.
   b List three things that are found in urine. (*Hint*: look back at Topic B2.18 if you need help.)

6 Phil eats some salted peanuts. Jill drinks a large glass of water.
   a Draw two flow charts to show how the change in balance of mineral ions and water in their bodies is brought back to normal.
   b Explain why this balance is important.

A steady temperature

By the end of this topic you should be able to:

- describe where, how and why temperature is monitored in the body
- explain how body temperature is controlled.

Different parts of our bodies can be at different temperatures, but the central **core body temperature** is usually kept steady at about 37 °C. The heart, liver, lungs and kidneys are major organs in the core of the body.

1 Look at photograph A.
   a The temperature of the background in the photo is blue. What does that tell you about the body temperature of the man?
   b Which areas of his body are hottest?
   c Which areas of his body are coolest?
   d Are the major organs in the hotter or cooler parts of the body?

2 Many reactions take place in cells in the major organs of the body. Think back to what you learnt earlier in this unit and suggest why the major organs need to be kept at a steady temperature.

To keep something steady, you have to **monitor** any changes. There must also be mechanisms for responding to any changes and getting back to the normal position. For example, a thermostat in a room monitors the air temperature. If the temperature falls too low, the thermostat will cause a radiator to switch on and heat the room. When the temperature gets too high, the thermostat will make the radiator switch off, and so the room cools. This kind of control is called **feedback control**.

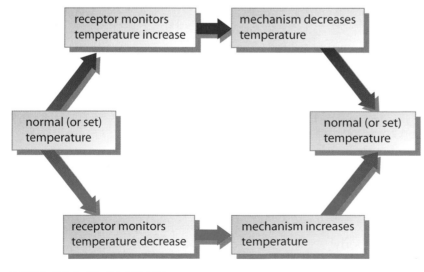

receptor monitors temperature increase → mechanism decreases temperature

normal (or set) temperature

normal (or set) temperature

receptor monitors temperature decrease → mechanism increases temperature

B Feedback control loop.

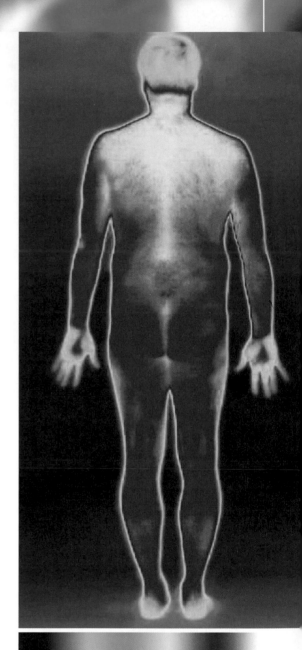

coldest   hottest

A This **thermogram** shows the temperature of different parts of the body.

3 Write a definition for the term feedback control.

4 Copy the boxes in diagram B. Rewrite the text to show how a thermostat keeps the temperature of a room constant at 20 °C.

In your body there are receptors in the **thermoregulatory centre** in your brain. These monitor internal temperature by measuring the temperature of blood flowing through the brain. Receptors in your skin measure external temperature. Information from all the receptors goes to the thermoregulatory centre, which then tells your body how to respond to any changes.

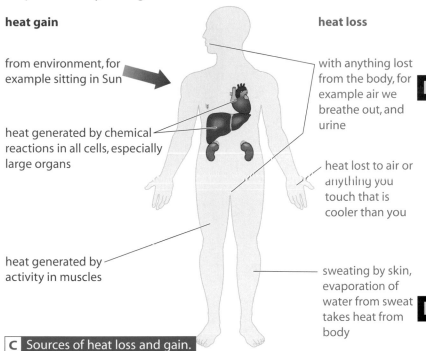

heat gain

from environment, for example sitting in Sun

heat generated by chemical reactions in all cells, especially large organs

heat generated by activity in muscles

heat loss

with anything lost from the body, for example air we breathe out, and urine

heat lost to air or anything you touch that is cooler than you

sweating by skin, evaporation of water from sweat takes heat from body

C Sources of heat loss and gain.

5 Copy diagram B again, but this time rewrite the text to show how your core body temperature stays constant at about 37°C.

6 Look at diagram C.
  a List three ways that your body can gain heat.
  b List three ways that your body loses heat.

H
7 Add notes to the diagram you drew in question **5** to include the mechanisms for controlling temperature.

8 a Jez is sweating. Give as many possible reasons for this observation as you can think of.
  b Suggest what Jez needs to do. Explain your answer.
H
  c Explain how changes in the blood vessels in his skin help to lose more heat from his body.

H

Core body temperature	Blood vessels in skin		Other changes
Too high	**dilate** (get wider)	skin surface — heat loss from skin — surface capillary — This increases flow of warm blood near skin, so heat can transfer easily to air. — deep skin blood vessel	• sweating releases liquid onto skin; sweat evaporates taking heat from skin • body hair lies flat
Too low	**constrict** (get narrower)	skin surface — little heat loss from skin — surface capillary — This keeps warm blood deeper in the skin so less heat is transferred to air. — deep skin blood vessel	• muscles 'shiver' – rapid contractions release heat energy from respiration • body hair raised, trapping layer of air next to skin as insulation

D Mechanisms controlled by the thermoregulatory centre to raise and lower body temperature.

Blood glucose and diabetes

By the end of this topic you should be able to:

- describe how blood glucose concentration is monitored and controlled
- state what diabetes is and say how it can be treated.

When you digest starchy or sugary foods, they are broken down to glucose and absorbed into your blood. If the glucose stayed in your blood it would be dangerous as the glucose causes water to move out of cells. The body has to tell cells to take glucose out of the blood so it can be used for respiration or stored until it is needed.

The pancreas monitors blood glucose concentration. If the concentration goes too high, the pancreas releases a **hormone** called **insulin** that tells cells to take glucose out of the blood.

1 Copy the top loop from diagram B in Topic B2.20. Write new text for the boxes to show how blood glucose concentration changes after eating and when insulin is released.

Blood glucose concentration

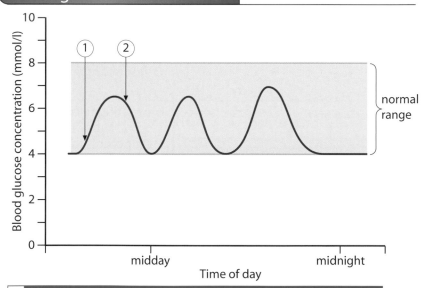

A Eating foods that contain sugar or starch soon raises the concentration of glucose in the blood.

B Blood glucose concentration normally changes during the day.

2 Look at graph B.
   a Which point, ① or ②, shows just after a meal was eaten? Explain your answer.
   b Which point, ① or ②, shows when insulin was released by the pancreas? Explain your answer.
   c Between which values does blood glucose concentration normally vary?

Diabetes and blood glucose concentration

Some people cannot control their blood glucose concentration properly. We say they have **diabetes**. Symptoms include excreting glucose in urine, unusual thirst and lack of energy.

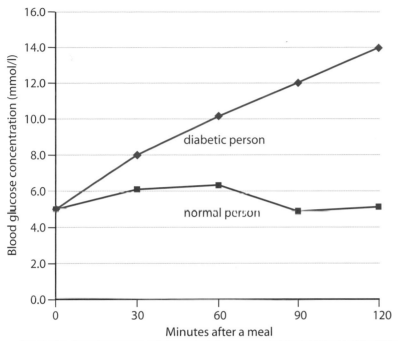

C Blood glucose concentrations of a normal person and a diabetic person after a meal.

If blood glucose concentration stays high, it can lead to coma and even death. So it is essential that it is brought back down again. There are different causes of diabetes and they are treated in different ways. As long as diabetics follow their treatment, they can lead normal lives.

Type of diabetes	Cause	Treatment
Type 1	pancreas doesn't release insulin	injections of insulin and healthy controlled diet
Type 2	pancreas makes insulin but body cells don't respond properly to it	for most: controlled diet and aerobic exercise for some: tablets to help body make more insulin or respond better to it in a few cases: injections of insulin

D Diabetes types.

4 Explain why diabetes must be treated.

5 a Suggest which part of the diet needs most careful control for diabetics. Explain your answer.
  b Why does aerobic exercise help control blood glucose concentration?

3 Look at graph C.
  a Describe the shape of the curve for the normal person.
  b Describe the shape of the curve for the diabetic person.
  c Explain the difference between the two curves.
  d How much higher has the blood glucose concentration of the diabetic increased beyond the normal range in your answer to question **2c**?

E Exercise uses glucose in respiration to make energy. Eluned Salisbury has diabetes and needs to be aware of the amount of exercise she does when controlling blood glucose concentration.

6 You have heard that a friend is worried because they have just been diagnosed as a diabetic. Write a fact sheet for them to explain:
  a what diabetes is
  b how it is caused
  c how it can be treated so that they can lead a normal life.

Body cell division

By the end of this topic you should be able to:

- explain that body cells have two sets of genetic information because their chromosomes are in pairs
- describe how body cells and offspring of asexual reproduction are produced by mitosis.

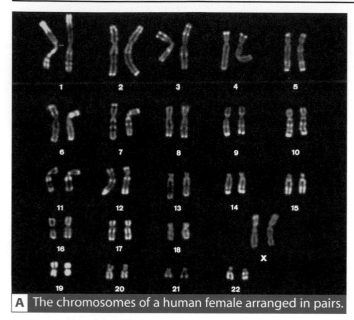

A The chromosomes of a human female arranged in pairs.

There are 46 **chromosomes** in the nucleus of almost every human body cell. If you took a photograph of the chromosomes from one cell, you could cut them out and arrange them as they are in photograph A.

The two chromosomes in each pair carry information for the same **characteristics**, such as eye colour. So each body cell carries two sets of information for all your characteristics. You inherited one set from your father and one set from your mother.

4 Explain why your chromosomes can be arranged in pairs.

5 Explain why some of your characteristics look like your mother, and some like your father.

Your body cells divide to make new cells when you grow, or when you repair damaged tissue. When the cells divide they need to produce identical copies of themselves, so that the new cells have all the information that the original cell had. So before a cell divides it needs to copy all that information. This means copying all the chromosomes. Then the chromosomes need to be carefully separated so that both new cells get one copy of each chromosome. This kind of division is called **mitosis**, and this is shown in diagram B.

1 How many pairs of chromosomes are there in most human body cells?

2 Where are the chromosomes found in a cell?

3 In photograph A, one pair of chromosomes is not numbered. These are chromosomes that show that the person the cell came from is female. How is this pair of chromosomes labelled for a human female?

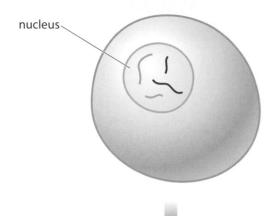
nucleus

This cell has one large pair and one small pair of chromosomes.

The chromosomes are copied.

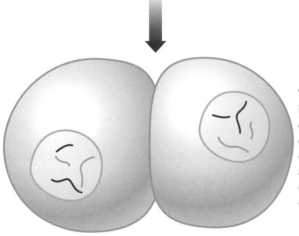
When the cell divides in two, each cell gets one new copy of each chromosome. So each cell has one large pair and one small pair of chromosomes. The cells are identical.

The chromosomes are drawn short here, and coloured, so it is easier to see what is happening. They don't really look like this.

B Mitosis.

In some species, mitosis can produce new individuals. Instead of staying together after division, the new cells (or groups of cells) separate and grow into new individuals. These offspring are all **clones** because they all have the same genetic information.

6 Why must everything in the cell be copied before the cell divides by mitosis?

7 How many new cells are made from the original cell in mitosis?

8 Suggest two situations when your body cells need to divide.

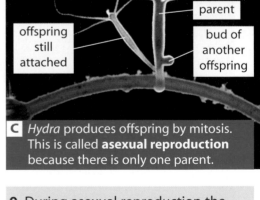
parent
offspring still attached
bud of another offspring

C *Hydra* produces offspring by mitosis. This is called **asexual reproduction** because there is only one parent.

9 During asexual reproduction the cells of the parent divide by mitosis.
   a Explain why this means that the offspring have exactly the same number of chromosomes as the parent.
   b Explain why this means the offspring have the same genetic information as the parent.

59

Different kinds of cells

> **By the end of this topic you should be able to:**
>
> - describe what cell differentiation is and when it happens in animals and plants
> - explain what stem cells are
> - give examples of how stem cells might be used to treat diseases
> - list some of the social and ethical issues about using embryonic stem cells.

A There are many kinds of cell in your body that do different jobs.

Your body contains many different kinds of cells, and almost all of them are **differentiated** with special shapes to do different jobs. When an animal egg cell is fertilised and starts dividing to make an **embryo**, the cells in the early stages can differentiate into any kind of cell. However, once cells are differentiated, they can only make more of the same kind of cell when they divide. These new cells can only be used to replace old cells or repair damage.

1 a Write a list of as many different kinds of cell in your body as you can think of.
   b Add to your list the job that each kind of cell does.

2 a Why can a muscle cell only divide to make more muscle cells?
   b Why do differentiated body cells divide to make more of the same kind of cell?

3 A cell in an early embryo can differentiate into any kind of cell. Explain why.

In plants, it is different. Many cells in many parts of a fully grown plant keep the ability to differentiate into any kind of cell.

4 Explain why we can grow whole new plants from leaf, root or shoot cuttings, but not a complete human from an arm or leg.

Undifferentiated animal cells are called **stem cells**. The cells in an early embryo are called 'embryonic stem cells'. There are also stem cells in tissues in your body, such as in your **bone marrow**. These are difficult to extract but, when they have been extracted, they can be treated to turn into many kinds of differentiated cells.

5 a Write down two sources of stem cells.
  b Suggest which would be the easier source to get stem cells from.
  c Explain your answer.

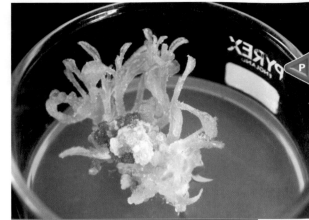

B In tissue culture, we take undifferentiated cells from the tip of a plant shoot, and treat them so that they develop into differentiated cells in leaves, shoots and roots.

Stem cells for treatment

People suffer from many diseases that are caused by cells not working properly or being damaged. There is a lot of research taking place into how embryonic stem cells could be used to make replacement cells for ones that are not working properly. For example, nerve cells grown from stem cells could cure **paralysis**, and insulin-secreting cells grown from stem cells could one day cure type 1 diabetes. However, some people argue that embryos have the potential to become whole beings and so should not be used like this.

C Tyson Gentry (an American footballer) was paralysed when an accident damaged nerve cells in his spine. Now he cannot walk.

6 a How could stem cells be useful in the future?
  b Suggest why some people argue against embryonic stem cell research.

7 A quote in a newspaper about a paralysed person says: 'Stem cells could make me walk again'. Explain how this could be done and why stem cells make it possible.

Making gametes

P

H **By the end of this topic you should be able to:**
- explain what gametes are and where they are made
- explain what meiosis is.

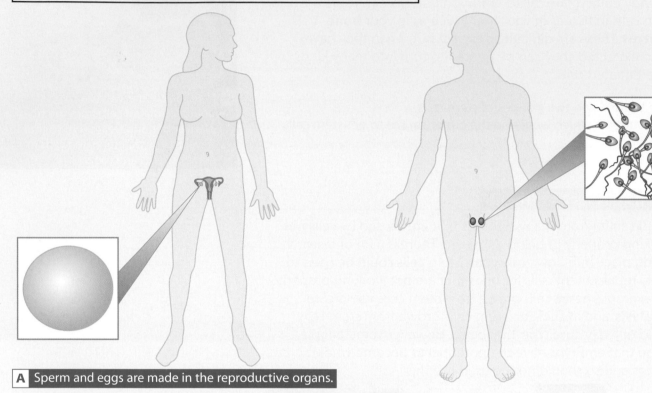

A Sperm and eggs are made in the reproductive organs.

Cells inside the **testes** and **ovaries** divide to produce **gametes**. Gametes are different to body cells because they only contain one set of chromosomes. The cells that divide to make them start with the same number of chromosomes as body cells.

P

1 In humans, what is the name of:
  a the female reproductive organs
  b the male reproductive organs
  c the female gamete
  d the male gamete?

2 Give one difference between the chromosomes in gamete cells and the chromosomes in body cells.

3 Suggest what must happen to the chromosomes during cell division to produce gametes.

The kind of cell division that makes gametes is called **meiosis**, and is shown in diagram B. At the end of meiosis, each cell has just one set of chromosomes in its nucleus. After division is complete, the cells develop the special features of either egg cells or sperm cells.

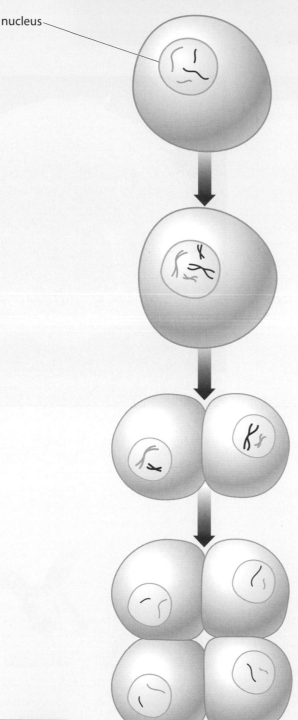

nucleus

This cell has one large pair and one small pair of chromosomes.

The chromosomes are copied.

The cell divides in two. Each new cell gets one chromosome from each pair but the copies of the chromosomes stay together. So each cell has two copies of one large chromosome and two copies of one small chromosome.

The cells divide in two again. Each new cell gets one copy of each chromosome. So each cell has one large and one small chromosome.

These cells are not identical.

B Meiosis.

4 a How many new cells are made from the original cell during meiosis?
  b How does this differ from the number of cells made during mitosis?

5 a Give two other differences between the cells produced by meiosis and by mitosis.
  b Explain how meiosis leads to variation between offspring.

6 Compare diagram B with diagram B in Topic B2.22.
  a Describe the similarities and differences between mitosis and meiosis.
  b Explain why the differences lead to the differences between body cells and gametes.

After fertilisation

By the end of this topic you should be able to:

- explain that gametes fuse to form a single cell and that the individual grows by mitosis
- explain why sexual reproduction leads to variation
- explain how sex is inherited from parents
- interpret a genetic diagram.

In a gamete, such as a sperm cell or egg cell, there is only one set of chromosomes. During fertilisation, one sperm cell **fuses** with one egg cell to form a new single cell that contains two sets of chromosomes. The cell then divides over and over again by mitosis. This forms an embryo that can eventually develop into a complete new individual.

1 **a** How many sets of chromosomes does:
   - **(i)** a sperm cell have
   - **(ii)** an egg cell have?
   b How many sets of chromosomes does the fertilised cell have?
   c How many sets of chromosomes will all the cells in the embryo have? Explain your answer.

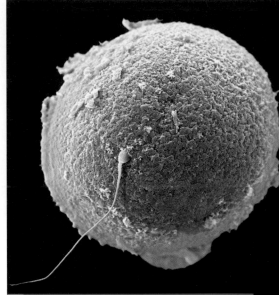

A Fertilisation of a human egg by one sperm cell.

As the cells produced by meiosis contain different sets of genetic information, this means that **sexual reproduction** produces offspring that are different from each other. We say they show **variation**.

B The offspring of these rabbits show variation.

One of the most obvious differences in offspring is that some are male and others are female. What makes someone male or female is dependent on the chromosomes they inherit in the gametes from their parents.

2 Explain why the offspring from sexual reproduction show variation.

3 Do the offspring of asexual reproduction show variation? Explain your answer.

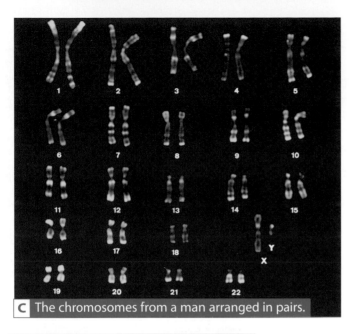

C The chromosomes from a man arranged in pairs.

4 Compare photograph C with photograph A in Topic B2.22. How are the chromosomes similar?

5 One pair of chromosomes, the **sex chromosomes**, is different.
 a Write down the sex chromosome pair for a man.
 b Write down the sex chromosome pair for a woman.

Interpreting genetic diagrams

Diagram D shows how being male or female is inherited. This is set out as a **genetic diagram**.

• At the top we show the parents and the chromosome pair that we are interested in from each of them.
• The middle of the diagram shows how the parent chromosomes separate into the gametes during meiosis.
• The bottom row shows the combination of each possible gamete from the father with each possible gamete from the mother.

We could have shown just one type of gamete for the mother, because the two possible gametes are the same. But it is a good habit to show all the possible gametes from meiosis and all the possible combinations.

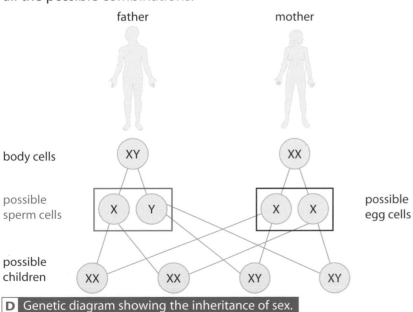

D Genetic diagram showing the inheritance of sex.

6 a Which type of sex chromosome does a woman produce?
 b How many kinds of sex chromosome does a man produce?
 c Is the chance that a child will be a boy or girl the same or different? Explain your answer.
 d Which parent has the gametes that control whether the children are boys or girls?

7 a Write a short presentation to explain how inheritance of sex is an example of variation due to sexual reproduction.
 b Explain why there would not be this variation if we reproduced asexually.

65

The genetic code

By the end of this topic you should be able to:

- describe how chromosomes made of DNA code for proteins
- describe how DNA fingerprinting can identify individuals.

Chromosomes are made of a chemical called **DNA** (deoxyribose nucleic acid). DNA has an unusual structure that we call a double helix. Imagine a ladder that is twisted round and round at the top – this makes the shape of DNA that you can see in photograph A.

Each chromosome that you have seen in earlier topics is one very long DNA molecule, with many thousands of 'rungs' in the double helix. Each chromosome is made up of thousands of **genes**. So each gene is a short piece of DNA.

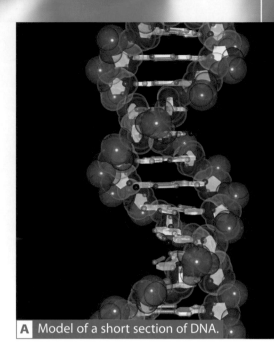

A Model of a short section of DNA.

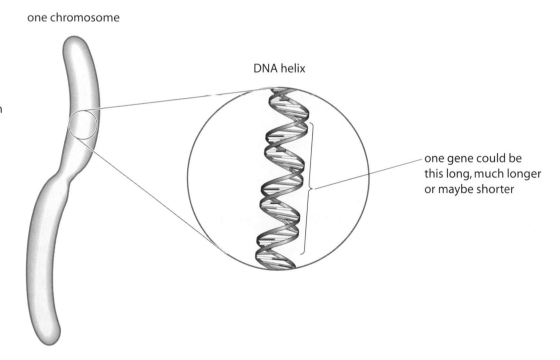

one chromosome

(Note that chromosomes are fat like this only during division. Then the DNA is highly folded. DNA is much narrower than this.)

DNA helix

one gene could be this long, much longer or maybe shorter

B Relationship between chromosomes, genes and DNA.

H You can't see where one gene ends and another starts by looking at the DNA. It all looks very similar. The only thing that varies along the DNA molecule is the rungs. There are four different chemicals, called **bases**, that can make the rungs. What is important is the order of bases along the DNA. This order is called the **genetic code**.

1 Describe a double helix.

2 Describe the relationship between a chromosome, DNA and a gene.

H Different groups of bases code for different amino acids, or say where a gene ends. The code also shows the order that the amino acids should be linked together to make a protein. Every protein has a different order of amino acids. When the cell needs to make a protein, the code for that protein is copied from the DNA and taken to another part of the cell where it is used to build the protein.

3 What are proteins made of?

4 Explain why the order of the bases on the DNA decides which protein is made.

5 Are all genes the same length? Explain your answer.

C Building a protein from the DNA's genetic code.

We can look at parts of someone's DNA using **DNA fingerprinting**. A small sample of DNA, such as from a sample of blood or semen, a hair or just a few skin cells is needed to do this. The DNA is split into short pieces. Then fluorescent dye is attached to bits of the DNA with particular orders of bases. Electrophoresis is used to separate the pieces of DNA, and the dye shows a pattern that is only likely to have come from one person.

D DNA fingerprints from many samples can be checked at the same time.

6 Identical twins have the same DNA fingerprints. Explain why.

7 Suggest how DNA fingerprinting could be used to:
  a identify the father of a baby
  b identify who carried out a crime
  c identify a dead body.
     (*Hint*: think about who samples will need to be taken from to compare in each case.)

H

8 a Explain why, unless you are an identical twin, your DNA fingerprint will be different to everyone else's.

  b Draw a flow chart to show how your DNA codes for the proteins in your body.

Different characteristics

By the end of this topic you should be able to:

- explain that different forms of a gene are called alleles and that these can produce different characteristics
- explain what dominant and recessive alleles are
- interpret genetic diagrams involving dominant and recessive alleles
- explain Mendel's ideas about inherited factors.

A Different alleles for the flower colour gene mean that some pea plants have white flowers and others have purple flowers.

Genes code for the building of proteins which produce our characteristics. For example, there is a gene in pea plants that controls the characteristic of flower colour. Not all pea plants have flowers of the same colour. This is because the plants have different forms of the flower-colour gene. The different forms are called **alleles**.

1 Write a definition for the word 'allele'.

2 Explain why some pea plants have purple flowers and some have white flowers.

Mendel's studies

Our understanding of the inheritance of characteristics began with the work of Gregor Mendel (1822–1884). Mendel saw that when he **crossed** purple-flower plants with white-flower plants the offspring all had purple flowers. He then crossed some of these first generation offspring with each other and found that in the next generation some plants had purple flowers, but some had white flowers. He wondered how white flowers could 'disappear' for a generation, but then come back.

We now understand how genes are inherited and can explain Mendel's observations. All pea-plant chromosomes, like ours, come in pairs. Each pair carries genes for the same characteristics. So there are two genes for flower colour. One gene comes from the father and one from the mother. Sometimes the forms, or alleles, of these genes are the same and sometimes they are different. The combination of alleles determines what the characteristic will look like.

In pea plants, we say that the purple-flower allele is **dominant**, and the white-flower allele is **recessive**. This is because when they are together you can only see the effect of the purple allele.

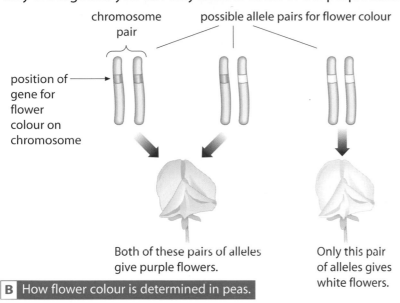

position of gene for flower colour on chromosome

chromosome pair

possible allele pairs for flower colour

Both of these pairs of alleles give purple flowers.

Only this pair of alleles gives white flowers.

B How flower colour is determined in peas.

Interpreting inheritance

We can show the inheritance of flower colour using a genetic diagram. We use a capital letter to stand for the dominant allele, and the same small letter for the recessive allele.

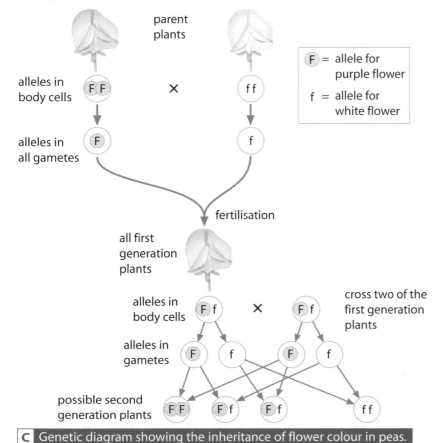

parent plants

alleles in body cells

F = allele for purple flower

f = allele for white flower

alleles in all gametes

fertilisation

all first generation plants

alleles in body cells

cross two of the first generation plants

alleles in gametes

possible second generation plants

C Genetic diagram showing the inheritance of flower colour in peas.

3 a Explain what we mean by a dominant allele.
b Explain what we mean by a recessive allele.

H 4 Explain why Mendel could cross purple-flower plants and get some offspring with white flowers.

5 Look at diagram C.
a What were the flower colours of the parent plants?
b What were the alleles in each of the parent plants?
c What colour were the flowers of all the first-generation offspring?
d What were the alleles in all the first-generation offspring?
e What possible combinations of alleles are there in the second-generation offspring?
f How often does each combination occur?
g What proportion of the second generation will have white flowers?

6 Copy diagram C.
a Add your own notes to explain what happens in each step.
b Complete your notes by explaining what we mean by a 'dominant allele' and a 'recessive allele'.

Inheriting diseases

Sometimes there is a change, or **mutation**, in an allele for a gene. This can change the protein that it makes, and if the protein doesn't work properly it will cause a disease, or disorder, that can be inherited.

An example of this is **Huntington's disease**. This is a disorder of the nervous system, which is caused by a protein from a mutated allele. The protein produces damage in the brain, which affects movement control, memory and mood. The faulty allele can be passed on to offspring during sexual reproduction, so the disease can be inherited.

1 Describe the differences in the brains in photograph A.

2 Can Huntington's disease be passed on:
a by someone sneezing over you
b by inheriting it from your parents?
Explain your answer.

Huntington's disease is caused by a dominant allele.

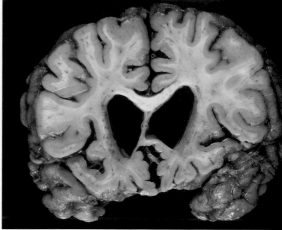
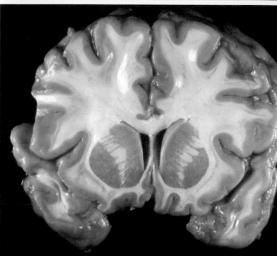

A The top picture shows the brain from someone with Huntington's disease. The bottom picture shows a normal brain.

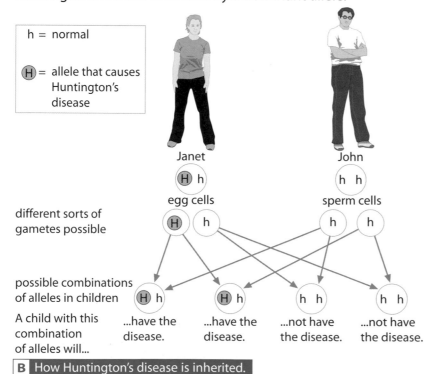

h = normal

(H) = allele that causes Huntington's disease

B How Huntington's disease is inherited.

3 Does a dominant allele mean that:
a you only see symptoms of the disease if two copies of the allele are inherited
b you see the effect if only one copy of the allele is inherited?
Explain your answer.

4 In diagram B, Janet has one copy of the allele that causes Huntington's disease. What are the chances of Janet and John's children inheriting the disease: no chance, 1 in 2, or 1 in 4?

Another disorder caused by a faulty allele is **cystic fibrosis**. This affects the secretion of mucus into parts of the body such as the lungs, gut and testes. Normally mucus is watery and flows easily. So, for example, it is easy to cough up mucus from the lungs. In cystic fibrosis the faulty version of the gene leaves mucus thick and sticky. The thick mucus causes problems.

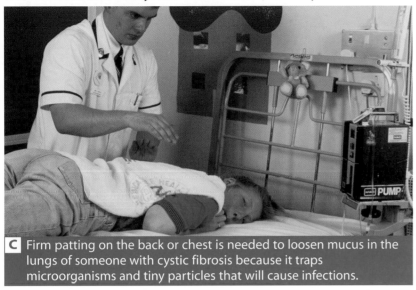

C Firm patting on the back or chest is needed to loosen mucus in the lungs of someone with cystic fibrosis because it traps microorganisms and tiny particles that will cause infections.

Cystic fibrosis is caused by a recessive allele, so someone only shows the effect of the disorder if they have two copies of the faulty allele. Someone who has one copy of the faulty allele is called a **carrier**. They don't show the symptoms but can pass the gene on to their children (see diagram D.)

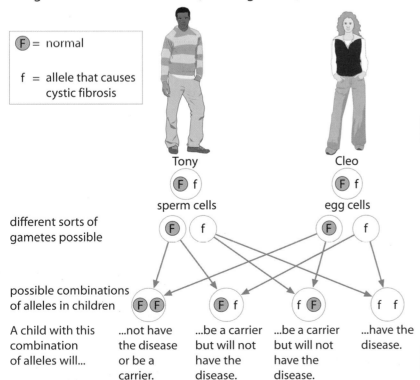

D How cystic fibrosis is inherited.

5 Look at diagram D.
   a Are Tony and Cleo carriers of the cystic fibrosis allele?
   b Will they show the symptoms of the disorder? Explain your answer.

H 6 What are the chances of Tony and Cleo having a child that will develop cystic fibrosis?

7 Use diagrams B and D to say how Huntington's disease and cystic fibrosis can be inherited. Your explanation should say:
   a if the faulty allele is dominant or recessive
   b how many parents must have the faulty allele for the offspring to develop the disorder.
   c what the chances are of the parents having a child that will develop the disorder.

Embryo screening

By the end of this topic you should be able to:

- describe how embryos can be screened for genetic diseases
- make informed judgements about issues concerning embryo screening.

Many genetic disorders can seriously affect the person's life. They can also affect how long the person is likely to live. Such people may need support and care from their families, and may need additional medicine and health care.

1 Explain why genetic disorders can be passed on to children.

2 Suggest why Oli doesn't want his children to inherit the disorder.

People with the faulty allele might also pass the disorder on to their children before they know they have it. For example, the symptoms of Huntington's disease often don't occur until between 30 and 50 years of age, and in cystic fibrosis both parents have to be carriers of the faulty allele for their child to suffer from the disease. People who think they might have a faulty allele can have tests to find out if they do.

A Oli has cystic fibrosis. He knows that he wouldn't want his children to suffer in the ways he does.

3 **a** Explain why a carrier of a genetic disease might not know that they are a carrier.
  b Explain why someone with Huntington's disease may pass the allele on to their children before they know they have the disease.

4 Imagine there is a history of Huntington's disease in your family.
  a Give one reason why you might want to know if you have the allele or not.
  b Suggest why you might *not* want to know you have the allele.

We know we are both carriers of cystic fibrosis so we have a 25% chance of passing it on to any child we have.

We could avoid the risk by not having any children or by adopting instead.

B

Making decisions

People who know they have faulty alleles have to make difficult decisions about having children. One way to make these decisions easier is **embryo screening**. This is when eggs are taken from the woman's ovaries and fertilised by sperm from her partner in a dish during IVF (*in vitro* fertilisation). The embryos are tested to see if they will suffer from the disease. Only an embryo that won't is placed in the womb of the woman to develop into a baby.

C Baby Roger Farre's embryo was screened to make sure he didn't have the allele for Huntington's disease.

Some people think this is good because it will reduce the suffering that families have to cope with. Other people feel that doing this changes how everyone thinks about those who have inherited disorders, and that we should value everyone equally no matter what they have in their genes.

Type of issue	Explanation
Ethical	whether we think something is right or wrong
Social	how something affects people, both individuals and all of society
Economic	where money is involved

D There are different kinds of issue we have to think about.

5 Explain how embryo screening might mean there are no inherited diseases in the future.

6 Suggest how embryo screening might affect how people respond to individuals with genetic disorders.

7 Should embryo screening be used to test for genetic diseases such as Huntington's disease or cystic fibrosis?
  a Use table D and the information on these pages to give examples of each kind of issue that people need to think about to answer this question properly.
  b Add notes to show how the view of a person who could pass on an allele for an inherited disorder might differ from someone who doesn't have the allele.

P
D
P
P

Investigative Skills Assessment 2

These questions refer to an experiment carried out by a student to investigate how the rate of reaction of the enzyme amylase is affected by temperature.

The student placed amylase in a starch solution. Samples of this solution were placed in water baths set at different temperatures.

Each sample was tested every 2 minutes for 20 minutes.

They tested each sample with iodine. The results for each sample are shown below. A tick (✔) indicates the presence of starch, a cross (✗) indicates that no starch was observed.

Time (minutes)	Temperature (10 °C)	Temperature (20 °C)	Temperature (40 °C)	Temperature (60 °C)	Temperature (80 °C)
0	✔	✔	✔	✔	✔
2	✔	✔	✔	✔	✔
4	✔	✔	✔	✔	✔
6	✔	✔	✔	✔	✔
8	✔	✔	✗	✔	✔
10	✔	✔	✗	✔	✔
12	✔	✔	✗	✔	✔
14	✗	✗	✗	✔	✔
16	✔	✗	✗	✗	✔
18	✗	✗	✗	✗	✔
20	✗	✗	✗	✗	✔

1 What is:
   a the dependent variable *(1 mark)*
   b the independent variable? *(1 mark)*

2 Describe what the student would have seen happening when testing each sample for starch in the 20 °C experiment. *(1 mark)*

3 Describe **one** difficulty that the method has for producing accurate results. *(1 mark)*

4 a Looking at the data, which result does not fit the pattern? *(1 mark)*
   b Explain what should be done with this result, and why, when analysing the data and forming a conclusion. *(1 mark)*

5 The student wanted to plot a graph of the results.
   a Which kind of graph would display those results best? *(1 mark)*
   b Explain your choice. *(1 mark)*

6 a Suggest **one** way in which you could obtain a more accurate record of the temperature that amylase works best at. *(1 mark)*
   b Explain your answer. *(1 mark)*

7 a Suggest **one** way the student could have made the results more reliable. *(1 mark)*
   b Explain your answer. *(1 mark)*

8 ✎ Describe **fully** the relationship between temperature and the rate of reaction of amylase, as shown by these results. Then write a conclusion based on the evidence from this investigation. *(4 marks)*

Total = 16 marks

Assessment exercises Foundation

1

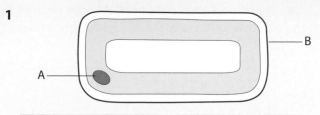

cell wall, cell membrane, nucleus, cytoplasm

Look at the diagram of the plant cell.
a Use the list to help you identify the two parts labelled A and B. (*1 mark*)
b How do you know that it is a plant cell? (*1 mark*)
c Name **two** other structures found only in a plant cell. (*2 marks*)

2

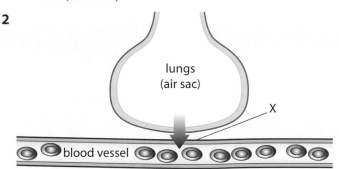

The diagram above shows a blood vessel that passes next to an air sac in the lungs.
The arrow shows the net movement of oxygen molecules.
a What is the process that describes the movement of oxygen across membrane X? (*1 mark*)
b Membrane X does not let large molecules through. What kind of membrane is this? (*1 mark*)
c Where is the concentration of oxygen highest? (*1 mark*)
d Explain your answer to **c**. (*2 marks*)

3 A scientist explains photosynthesis to a farmer. To help explain what happens he writes the equation below.

carbon dioxide + water $\xrightarrow{\text{light energy}}$ glucose + oxygen

a Where does the carbon dioxide come from? (*1 mark*)
b If the amount of carbon dioxide was increased what would happen to the amount of glucose? (*1 mark*)
c What would happen to a plant if it didn't get enough light? (*1 mark*)
d Explain why this would happen. (*1 mark*)

4 Study the flow diagram of the carbon cycle.

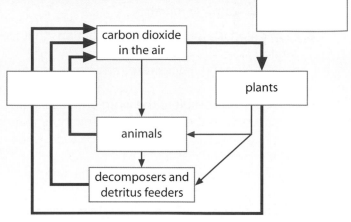

a Copy and complete the diagram by filling in the two empty boxes. (*2 marks*)
b What do decomposers and detritus feeders do? (*1 mark*)
c Give the name of **one** kind of decomposer. (*1 mark*)

5 A shopper is trying to make a choice between buying free-range pork chops or ordinary pork chops.
a The pigs which produced the ordinary pork chops were kept in small pens. Suggest why the farmer did this. (*1 mark* **HSW**)
b The free-range pork chops are more expensive. Why might the shopper buy them? (*1 mark* **HSW**)

6 Mary measures the amount of oxygen produced by pondweed at different light levels. A graph of her results is shown below.

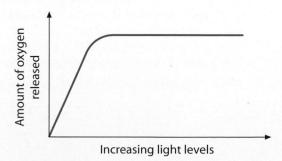

a Which is the dependent variable? (*1 mark* **HSW**)
b Are the two variables discrete or continuous? (*1 mark* **HSW**)
c What happens to the amount of oxygen released as the light level increases? (*1 mark* **HSW**)

7 a How do catalysts affect a chemical reaction? (*1 mark*)

b What is the name given to biological catalysts? (*1 mark*)

c Which kind of molecules are biological catalysts made from? (*1 mark*)

8 Respiration is a set of chemical reactions that use catalysts.

a Where in a cell does respiration take place? (*1 mark*)

b Copy and complete the word equation that summarises the reactions in respiration: (*1 mark*)
glucose + _____ ➔ carbon dioxide + water + energy

c Give **one** way in which the energy released during respiration is used in animals. (*1 mark*)

d Give **one** way in which the energy released in respiration is used in plants. (*1 mark*)

9 Ahmed has diabetes. His body does not make enough insulin.

a Where is insulin normally made? (*1 mark*)

b What does insulin do in Ahmed's body? (*1 mark*)

c Give **one** way Ahmed can control the symptoms of diabetes. (*1 mark*)

10 Shirley has cystic fibrosis. Cystic fibrosis is caused by a recessive gene.

a What do we mean by a recessive gene? (*1 mark*)

b Explain how Shirley inherited cystic fibrosis. (*2 marks*)

c Shirley's parents could have detected the disease if they had chosen embryo screening. Why do some people object to embryo screening? (*2 marks **HSW***)

11 a What is the name of the long molecule that makes up our chromosomes? (*1 mark*)

b What are small sections of our chromosomes called that select for features such as hair colour, eye colour etc? (*1 mark*)

12 Acme Pharmaceuticals are looking to design a washing powder that will remove stains from clothes.

a Name an enzyme that will remove each of the following types of stain. Choose from the list below:

| lipase, protease, carbohydrase, isomerase |

i fat stains (*1 mark*)
ii protein stains (*1 mark*)

b Explain briefly how these enzymes work. (*1 mark*)

13 The graph below shows the action of an enzyme.

a What are the two variables shown on the graph? (*1 mark **HSW***)

b Which variable is the dependent variable? (*1 mark **HSW***)

c At what temperature does the enzyme work best? (*1 mark **HSW***)

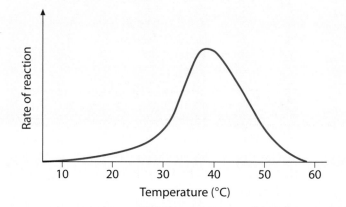

Assessment exercises Higher

1

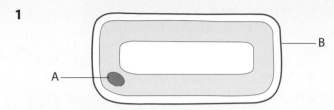

Look at the diagram of the plant cell.
a Identify the two parts labelled A and B. (*1 mark*)
b Explain the function of B. (*1 mark*)
c How do you know that it is a plant cell? (*1 mark*)

2

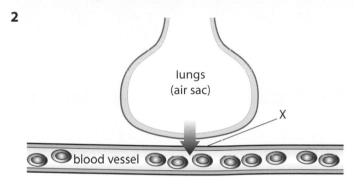

The diagram above shows a blood vessel that passes next to an air sac in the lungs.
a What is the name of the process that describes the movement of oxygen across membrane X? (*1 mark*)
b Membrane X does not let large molecules through. What kind of membrane is this? (*1 mark*)
c Explain why net movement of oxygen is from the air sac to the blood and not the other way. (*2 marks*)

3 A scientist explains photosynthesis to a farmer. To help explain what happens he writes the equation for photosynthesis.

$$6CO_2 + \underline{\quad\quad} \rightarrow \underline{\quad\quad} + 6O_2$$

a Copy and complete the equation. (*1 mark*)
b Where does the carbon dioxide come from? (*1 mark*)
c If the amount of carbon dioxide was increased what would happen to the amount of glucose produced? (*1 mark*)
d Give **one** other way the farmer might increase the amount of glucose made in his crops. (*1 mark*)
e Explain why increasing rate of photosynthesis increases crop yield. (*1 mark*)

4 Study the flow diagram of the partially completed carbon cycle.

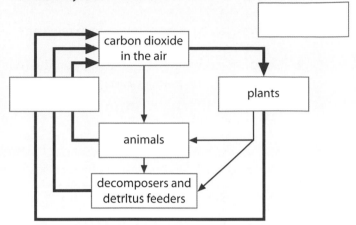

a Copy and complete the diagram by filling in the empty boxes. (*1 mark*)
b Explain how each process in **a** contributes to the carbon cycle. (*2 marks*)
c What do decomposers and detritus feeders do? (*1 mark*)
d Explain the importance of these organisms in the carbon cycle. (*1 mark*)

5 A shopper is trying to make a choice between buying free-range pork chops or ordinary pork chops.
a The pigs which produced the ordinary pork chops were kept in small pens. Suggest why the farmer did this. (*1 mark* **HSW**)
b The free-range pork chops are more expensive. Why might the shopper buy them? (*1 mark* **HSW**)

6 Mary measures the amount of oxygen produced by pondweed at different light levels. A graph of her results is shown below.

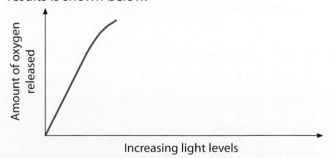

a Which is **(i)** the dependent variable, **(ii)** the independent variable? (*1 mark* **HSW**)
b Are the two variables discrete or continuous? (*1 mark* **HSW**)
c Copy and complete the graph to show what would happen as light levels were increased further. (*1 mark*)

7 a Which molecules are enzymes made from? (*1 mark*)

b Explain why the shape of an enzyme is important. (*2 marks*)

8 Respiration is a set of chemical reactions that use catalysts.

a Where in a cell does respiration take place? (*1 mark*)

b Copy and complete the symbol equation that summarises the reactions in respiration: (*1 mark*)

$$C_6H_{12}O_6 + \underline{\hspace{1cm}} \rightarrow \underline{\hspace{1cm}} + 6H_2O + \text{energy}$$

c Give **two** ways in which the energy released during respiration is used in animals. (*1 mark*)

d Give **two** ways in which the energy released in respiration is used in plants. (*1 mark*)

9 Ahmed has diabetes.

a Which hormone is his body not making enough of? (*1 mark*)

b Where is this hormone made? (*1 mark*)

c How does this hormone prevent diabetes? (*1 mark*)

d Identify **two** ways by which Ahmed can control the symptoms of diabetes. (*1 mark*)

10 Shirley has cystic fibrosis.

a Is the allele for cystic fibrosis dominant or recessive? (*1 mark*)

b Explain how Shirley inherited cystic fibrosis. (*2 marks*)

c How might Shirley's parents have detected the disease in Shirley before she was born? (*1 mark*)

d Why might some people object to your answer to **c**? (*1 mark* **HSW**)

11 Acme Pharmaceuticals are looking to design a washing powder that will remove stains from clothes.

a Name a type of enzyme that will remove each type of stain.

i fat stains (*1 mark*)
ii protein stains (*1 mark*)

b Explain briefly how these enzymes work. (*1 mark*)

12 The graph below shows the action of an enzyme measured at different temperatures.

a Which variable is the dependent variable? (*1 mark* **HSW**)

b How would each variable be measured? (*1 mark* **HSW**)

c What kind of variable is the independent variable? (*1 mark* **HSW**)

d How could you show that the graph was accurate? (*1 mark* **HSW**)

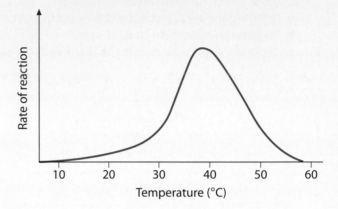

Glossary

aerobic respiration The breakdown of glucose using oxygen from the air to release energy, carbon dioxide and water.

allele One form of a gene.

allergic reaction A reaction, such as a skin rash, by the body to a substance.

amino acid The building block of proteins.

amylase Enzyme that catalyses the digestion of starch.

asexual reproduction Producing offspring without fertilisation, by mitosis.

H base (genetics) A molecule in DNA that is part of the genetic code.

biomass The total weight of living material that exists at a certain level in a food chain.

bone marrow The soft tissue inside the long bones of the body where some stem cells are found.

by-product Product formed in the process of making something else.

carbohydrase Enzyme that catalyses the breakdown of starch syrup to sugar syrup.

carbon cycle The way in which carbon atoms circulate from one organism to another in the environment.

carrier Someone who has one recessive allele for a genetic disorder, so they do not have the disease but can pass it on to their children.

catalyst Something that speeds up a reaction, but is not itself used up.

cell membrane The thin outer layer of a cell that controls what goes into and comes out of a cell.

cell sap A liquid that contains dissolved sugars and salts and is found in plant cell vacuoles.

cell wall Outer covering layer of a plant cell that provides support for a plant cell.

characteristic A feature. What a gene codes for.

chlorophyll Green pigment found in chloroplasts. It is used to trap light for photosynthesis.

chloroplasts Organelles in plant cells that contain chlorophyll.

chromosome Long length of DNA. Found in the nucleus of cells.

clone Offspring identical to the parent.

community All the living organisms present in an ecosystem.

concentration Measure of how much solute is dissolved in a solvent such as water.

concentration gradient The graduated difference in concentration between two different concentrations of a substance in two places.

H constrict Get narrower.

core body temperature Temperature deep inside the body.

cross (genetics) Mating one individual with another.

cystic fibrosis Genetic disease caused by a recessive allele. It makes mucus thick and sticky which causes many problems in the body.

cytoplasm The contents of a cell outside its nucleus, where the chemical reactions take place.

decomposers Organisms that decompose materials.

deficiency symptoms Changes in appearance caused by the lack of a particular mineral ion.

dehydration Containing too little water.

detergent Chemical that helps break down stains.

detritus Fragments of partly broken down plant or animal tissue.

detritus feeder Organism that feeds on detritus.

diabetes Disorder where the body cannot control the blood glucose concentration properly.

differentiated (cell) A cell that has become a particular type, such as a muscle cell.

diffusion The movement of particles through another substance from an area of high concentration to an area of lower concentration.

digest Break down into smaller molecules, such as in the gut.

H dilate Get wider.

DNA Deoxyribose nucleic acid. Chemical that makes up genes and chromosomes.

DNA fingerprinting Making an image of part of the DNA code.

dominant An allele which produces the characteristic when only one allele is present.

efficient (ecology) Reducing the amount of energy lost from the food chain.

embryo The early stages of development after an egg is fertilised.

embryo screening Checking embryos to see if they have genetic diseases or other characteristics.

energy store A deposit of starch in plants or fat in animals.

enzymes Chemicals that control the rate of reaction in cells.

excrete Remove from the body.

factory farming A system of large scale, industrialised and intensive agriculture. Animals raised are kept indoors and allowed little movement.

fatty acid One of the sub-units that, with glycerol, makes up fats and oils.

feedback control Where the result of a process or reaction is monitored and used to control the rate of the process or reaction.

fertilisers Chemicals that contain the ions that plants need so that they can grow properly.

food chain A pathway that energy and nutrients follow in a community.

fructose What glucose is changed to by an isomerase enzyme.

fuse (biology) Join together.

H **gamete** A sex cell.

gene A short piece of chromosome that codes for a characteristic or protein.

H **genetic code** The sequence of bases on a DNA molecule.

genetic diagram Diagram to show the inheritance of alleles and characteristics.

gland A group of cells responsible for the production and release of a particular substance.

glycerol One of the sub-units that, with fatty acids, makes up fats and oils.

homeostasis Keeping everything in balance in the body.

hormone Protein that tells some cells what to do.

Huntington's disease Genetic disorder caused by a dominant allele. A disease that affects the nervous system later in life.

insoluble Substance that does not dissolve in a liquid such as water.

insulin Hormone that controls blood glucose concentration.

isomerase Enzyme that catalyses the change of glucose to fructose.

limited The point beyond which a reaction does not increase.

limiting factor An environmental variable that slows or limits the rate of a reaction.

lipase Enzyme that catalyses the digestion of lipid.

lipid Another name for fat or oil.

H **meiosis** Division of cells to make gametes. The cells produced have only one set of chromosomes.

microorganism Organism which is too small to be seen without a microscope.

mineral ions Charged particles that plants absorb from the soil for healthy growth.

mitochondria Organelles that provide energy for a cell during respiration.

mitosis Division of a body cell that produces two identical cells.

monitor Keep checking.

mutation A change in a gene.

net movement The overall amount and direction of movement of particles during diffusion.

nucleus (biology) Large organelle that controls the activities of a cell.

organelles Structures found within a cell, with a particular function.

organs Have a characteristic shape, a specific function and are made up from different types of tissue.

osmosis Net movement of water molecules from a dilute to a more concentrated solution across a partially permeable membrane.

H **ovaries** The female reproductive organs in humans and other animals.

overhydration Absorbing too much water.

palisade cells Cells in the leaf involved in photosynthesis.

paralysis Unable to move.

partially permeable membrane Thin layer with tiny holes in it that allows some particles to pass through.

photosynthesis Process by which green plant cells produce glucose and oxygen from carbon dioxide and water using light energy.

protease Enzyme that catalyses the digestion of protein.

pyramid of biomass Diagram that shows the amount of biomass at each level in a food chain.

rate of photosynthesis The amount of photosynthesis occurring in unit time.

recessive An allele which produces the characteristic only when two of the alleles are present.

respiration The breakdown of sugar in cells to release energy, carbon dioxide and water.

ribosomes Tiny organelles where protein synthesis occurs inside a cell.

root hair cells Cells that have root hairs on them.

scavenger Animal that feeds on dead plants and animals.

sex chromosomes Chromosomes that determine whether the individual is male or female.

sexual reproduction Offspring produced by the fusion of a male and a female gamete.

soluble Substance that dissolves in a liquid such as water.

specialised cells Cells that have special features that are related to what they do. Also called differentiated cells.

stem cell Cell that can differentiate into many different types.

H **testes** The male reproductive organs in humans and other animals.

thermogram Diagram showing temperature.

thermoregulatory centre Part of the body which monitors and controls internal body temperature.

tissue A mass of similar cells.

toxic Poisonous.

urea Substance made in the liver from the breakdown of amino acids.

vacuole Space in the middle of a plant cell where cell sap is found.

variation Difference between individuals.

waste product A product of a reaction that isn't used for anything else.

xylem vessels Dead cells that transport water and mineral ions in plants.

yield (biology) Amount of a crop that a plant produces.

Structure and bonding

A Table salt.

B INEOS Chlor plant at Runcorn.

C Many useful substances are produced from chemicals made by the electrolysis of salt solution.

You use salt to flavour your food. However, salt is also an important raw material. Many useful substances are made from the chemicals produced by the electrolysis of salt solution. These substances include margarine, fertiliser, bleach, soap, disinfectant, PVC and paper.

By the end of this unit, you should be able to:

- describe the structure of atoms
- explain the difference between elements and compounds
- describe the three different types of bonding
- list the five different types of structure
- explain how the properties and uses of a substance depend on its structure
- describe what smart materials and nanomaterials are and what they can be used for
- **H** calculate how much of a chemical is needed for or made in a reaction
- calculate the empirical formula of a compound
- explain what is meant by atom economy and why yield is never 100% in a chemical reaction
- **H** calculate the percentage yield and atom economy for a reaction
- describe and explain what happens in electrolysis
- list some uses and useful products of electrolysis.

1 a Chemicals produced from the electrolysis of salt solution can be used to make other substances. Make a list of these.
   b Write a few sentences to describe how your life might be different without these substances.

The structure of atoms

By the end of this topic you should be able to:

- describe atoms using the terms 'protons', 'neutrons' and 'electrons'
- describe the relative charge and mass of protons, neutrons and electrons
- explain what isotopes are.

Atoms are made of three smaller sub-atomic particles called **protons**, **neutrons** and **electrons**. Table A shows the relative mass and electric charge of these particles. They are measured relative to a proton.

	Sub-atomic particle		
	Proton	**Neutron**	**Electron**
Relative charge	+1	0 (neutral)	−1
Relative mass	1	1	0.0005 (negligible)

A Relative mass and electric charge of sub atomic particles.

At the centre of the atom is a tiny **nucleus**. The nucleus contains protons and neutrons. Most of the mass of the atom is in the nucleus. The electrons orbit the nucleus in **energy levels** (**shells**).

Atoms are neutral because they contain the same number of negatively charged electrons and positively charged protons.

1 a Name the three types of particle inside atoms.
   b Which of these particles are inside the nucleus?

2 Atoms contain both positive and negative particles, but are neutral overall. Explain why atoms are neutral.

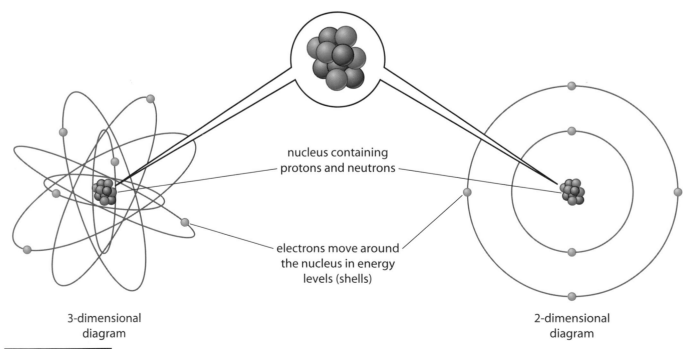

nucleus containing protons and neutrons

electrons move around the nucleus in energy levels (shells)

3-dimensional diagram

2-dimensional diagram

B Inside an atom.

Atoms of different elements have different numbers of protons, neutrons and electrons. Atoms are described by their **mass number** and **atomic number** (**proton number**) which can be shown as in diagram C.

Mass number = number of protons + number of neutrons ⟶ 23

Na

Atomic number (proton number) = number of protons ⟶ 11

| C | Mass number and atomic number. |

These two numbers can be used to work out how many protons, neutrons and electrons there are in an atom. Remember that atoms are neutral so the number of electrons always equals the number of protons.

For example, in $^{23}_{11}$Na, number of:

protons   = atomic number   = 11
neutrons = mass number – atomic number = 23 – 11 = 12
electrons = atomic number   = 11

Isotopes are atoms of the same element with a different mass number. In other words, they are atoms with the same number of protons, but a different number of neutrons. For example, there are two isotopes of chlorine.

Isotope	Protons	Neutrons	Electrons
$^{37}_{17}$Cl	17	20	17
$^{35}_{17}$Cl	17	18	17

| D | Isotopes of chlorine. |

The number of protons in the nucleus of an atom determines which element the atom is. All atoms with 17 protons are chlorine atoms. This is what makes both of the atoms in table D chlorine atoms. Atoms of different elements have different numbers of protons.

5 There are two common isotopes of bromine. They are shown below.

$^{79}_{35}$Br and $^{81}_{35}$Br

  a What are isotopes?
  b Draw a table to show how many protons, neutrons and electrons there are in each bromine atom.
  c Explain why both atoms are bromine atoms.

3 An atom of fluorine has 9 protons and 10 neutrons.
  a What is the atomic number of this atom?
  b What is the mass number of this atom?
  c How many electrons are in this atom? Explain your answer.

4 An atom has the atomic number 18 and mass number 40.
  a How many protons, neutrons and electrons does it contain?
  b What element is this atom? You will need to look at a Periodic Table to help you.

6 Two isotopes of carbon are shown below.

$^{12}_{6}$C and $^{14}_{6}$C

  a Sketch a diagram showing the structure of these two atoms.
  b Explain why they are isotopes.

Electronic structure

By the end of this topic you should be able to:

- describe how electrons are arranged in energy levels for the first 20 elements
- explain how elements are arranged in the Periodic Table
- relate the group that an element is in to the number of electrons in its highest energy level.

The elements in the Periodic Table are arranged in order of atomic number. Atomic number is the number of protons in the nucleus of an atom.

																	0
									H 1								**He** 2
1	**2**										**3**	**4**	**5**	**6**	**7**		
Li 3	**Be** 4										**B** 5	**C** 6	**N** 7	**O** 8	**F** 9	**Ne** 10
Na 11	**Mg** 12				transition metals						**Al** 13	**Si** 14	**P** 15	**S** 16	**Cl** 17	**Ar** 18
K 19	**Ca** 20	**Sc** 21	**Ti** 22	**V** 23	**Cr** 24	**Mn** 25	**Fe** 26	**Co** 27	**Ni** 28	**Cu** 29	**Zn** 30	**Ga** 31	**Ge** 32	**As** 33	**Se** 34	**Br** 35	**Kr** 36
Rb 37	**Sr** 38	**Y** 39	**Zr** 40	**Nb** 41	**Mo** 42	**Tc** 43	**Ru** 44	**Rh** 45	**Pd** 46	**Ag** 47	**Cd** 48	**In** 49	**Sn** 50	**Sb** 51	**Te** 52	**I** 53	**Xe** 54
Cs 55	**Ba** 56	**La** 57	**Hf** 72	**Ta** 73	**W** 74	**Re** 75	**Os** 76	**Ir** 77	**Pt** 78	**Au** 79	**Hg** 80	**Tl** 81	**Pb** 82	**Bi** 83	**Po** 84	**At** 85	**Rn** 86
Fr 87	**Ra** 88	**Ac** 89	**Rf** 104	**Db** 105	**Sg** 106	**Bh** 107	**Hs** 108	**Mt** 109	**Ds** 110	**Rg** 111							

A The Periodic Table.

1 In what order are the elements arranged in the Periodic Table?

2 What is the atomic number of an atom?

The electrons are arranged in energy levels (shells). Each energy level can hold a certain number of electrons. You need to know how the first 20 electrons are arranged. Two electrons can fit in the first energy level and eight electrons in the second energy level. The next eight electrons are in the third energy level and the next two electrons are in the fourth energy level.

Electrons always occupy the lowest available energy level. The lowest available energy level is the one closest to the nucleus. Electrons in each complete energy level are arranged in pairs.

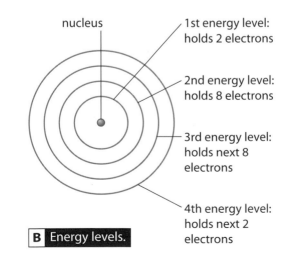

nucleus

1st energy level: holds 2 electrons

2nd energy level: holds 8 electrons

3rd energy level: holds next 8 electrons

4th energy level: holds next 2 electrons

B Energy levels.

The electronic structure can be drawn on a diagram or written in numbers. For example, the electronic structure of an aluminium atom is 2,8,3. This means there are two electrons in the first energy level, eight electrons in the second energy level and three electrons in the third energy level.

helium (He)	fluorine (F)	aluminium (Al)	calcium (Ca)
2 electrons	9 electrons	13 electrons	20 electrons
2	2,7	2,8,3	2,8,8,2

C Electronic structures of some atoms.

There is a link between the position of an element in the Periodic Table and its electronic structure. Elements in the same group have the same number of electrons in their highest energy level (their outer shell). For example, all the elements in Group 1 have one electron in the highest energy level; all the elements in Group 7 have seven electrons in the highest energy level.

3 The electronic structure of chlorine is 2,8,7.
   a What does this mean?
   b How many electrons are there in a chlorine atom?

4 An atom has 19 electrons.
   a Write its electronic structure.
   b Draw a diagram to show how the electrons are arranged.
   c Name this element.

1	2															3	4	5	6	7	0
							H 1 1														He 2 2
Li 3 2,1	Be 4 2,2															B 5 2,3	C 6 2,4	N 7 2,5	O 8 2,6	F 9 2,7	Ne 10 2,8
Na 11 2,8,1	Mg 12 2,8,2															Al 13 2,8,3	Si 14 2,8,4	P 15 2,8,5	S 16 2,8,6	Cl 17 2,8,7	Ar 18 2,8,8
K 19 2,8,8,1	Ca 20 2,8,8,2																				

D Electronic structures of the first 20 elements.

All the elements in the same group of the Periodic Table have similar chemical properties. This is because they have the same number of electrons in their highest energy level (in their outer shell).

5 Bromine has seven electrons in its highest energy level. Which group is bromine in?

6 Lead is in Group 4 of the Periodic Table. How many electrons are in its highest energy level (outer shell)?

7 Sulfur has the electronic structure 2,8,6.
   a Explain what this means. You should include a diagram of the electronic structure in your answer.
   b What does this structure tell you about the position of sulfur in the Periodic Table?

Elements and compounds

By the end of this topic you should be able to:

- explain the difference between elements and compounds
- identify whether a substance is an element or compound.

Hydrogen and chlorine are elements.

Sodium chloride and sodium hydroxide are compounds.

A Elements and compounds in the electrolysis of salt water.

There are just over 100 different types of atom. An **element** is a substance made from only one type of atom. This means that there are just over 100 different elements. They are listed in the Periodic Table (see page 225). An element cannot be broken down into simpler substances.

Hydrogen and oxygen are both elements. Hydrogen is a colourless gas that is flammable. Oxygen is a colourless gas in which other substances burn well. If you mix hydrogen and oxygen, then you still have the same substances: hydrogen and oxygen.

If a flame is put near the mixture of hydrogen and oxygen, a chemical reaction takes place producing the compound water. This is very different to both hydrogen and oxygen. It is a colourless liquid that puts certain types of fires out.

1 What is an element?

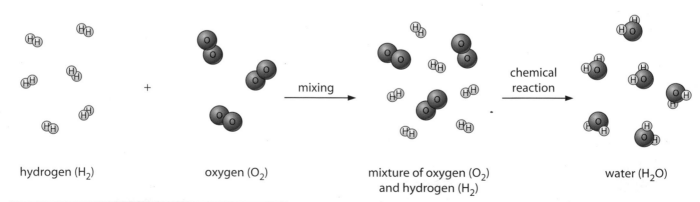

hydrogen (H_2)   +   oxygen (O_2)   mixing   mixture of oxygen (O_2) and hydrogen (H_2)   chemical reaction   water (H_2O)

B The reaction between hydrogen and oxygen.

In a mixture of hydrogen and oxygen, atoms of oxygen are not joined to atoms of hydrogen, they are just mixed together. In the compound water, the hydrogen and oxygen atoms are joined to each other.

The equation for this chemical reaction is:

hydrogen + oxygen $\rightarrow$ water

$$2H_2 + O_2 \rightarrow 2H_2O$$

A **compound** is a substance that contains atoms of more than one type chemically joined together. In other words, it is a substance made from different elements chemically joined together. Although there are only just over 100 elements, there are millions of known compounds.

The properties of each substance in a mixture are the same as before they were mixed. However, the properties of a compound are different to the elements from which it is made. For example, sodium chloride (common salt) is made from sodium and chlorine. Sodium and chlorine are both very dangerous, toxic elements. Sodium chloride is completely different to both sodium and chlorine. Sodium chloride is safe enough to eat.

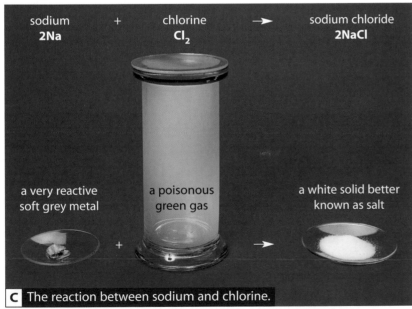

sodium 2Na + chlorine Cl$_2$ → sodium chloride 2NaCl

a very reactive soft grey metal + a poisonous green gas → a white solid better known as salt

C The reaction between sodium and chlorine.

Compounds can be broken down (decomposed) into simpler substances. This can often be done using heat (**thermal decomposition**) or electricity (**electrolysis**). For example, passing an electric current through molten aluminium oxide breaks the compound down into aluminium and oxygen.

There are also many common mixtures. For example, air is a mixture of several different elements and compounds. Rocks are also mixtures of several different substances. In a mixture, the separate substances are not chemically joined to each other and so have their own separate properties.

2 What is a compound?

3 What is the difference between a mixture of hydrogen and oxygen and the compound water. Explain in terms of how the atoms are arranged.

4 Some substances burn well in oxygen. Water molecules contain oxygen, but substances do not burn in water. Explain why.

5 Here is a list of substances: Mg, SO$_2$, Co, CO, Br$_2$, KBr, CaCO$_3$, K$_2$O.
a Which of these substances are elements?
b Which of these substances are compounds?

6 For each of the particles in diagram D, decide whether the particles represent the particles in an element, compound or mixture.

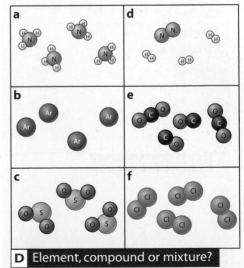

D Element, compound or mixture?

7 Carbon dioxide is a compound made from the elements carbon and oxygen. Use these three substances to explain the difference between elements and compounds. Include a diagram in your answer.

Ions

By the end of this topic you should be able to:

- describe what an ion is
- write or draw the electronic structure of ions and link them to the electronic structure of Group 0 elements
- explain how ions can be formed when metals react with non-metals.

The particles in many substances are **ions**. Ions contain a different number of protons to the number of electrons. Protons have a positive electric charge and electrons have a negative electric charge. This means that ions are electrically charged. Ions can be positively or negatively charged. Some common ions are shown in table A.

1 What is an ion?

2 Explain why ions are electrically charged.

3 Sulfur ions contain 16 protons and 18 electrons. What is the electrical charge of a sulfur ion?

4 Caesium ions contain 55 protons and 54 electrons. What is the electrical charge of a caesium ion?

	Sodium ion	Calcium ion	Aluminium ion	Chloride ion	Oxide ion
Symbol	Na^+	Ca^{2+}	Al^{3+}	Cl^-	O^{2-}
Number of protons	11	20	13	17	8
Number of electrons	10	18	10	18	10
Electrical charge	protons +11 electrons −10	protons +20 electrons −18	protons +13 electrons −10	protons +17 electrons −18	protons +8 electrons −10
Overall electrical charge	+1	+2	+3	−1	−2
Electronic structure	$[2,8]^+$	$[2,8,8]^{2+}$	$[2,8]^{3+}$	$[2,8,8]^-$	$[2,8]^{2-}$

A Some common ions.

B The electronic structure of some common ions and Group 0 atoms.

Common ions have electronic structures in which all the energy levels (shells) containing electrons are full. There are no partially filled energy levels. This means that ions have the same electronic structure as the Group 0 elements (**noble gases**). However, the hydrogen ion (H^+) contains no electrons at all.

Ions can be formed when metals react with non-metals. Metal atoms only need to lose a small number of electrons in order to have a full outer shell. Metals form positively charged ions. Many non-metal atoms only need to gain a small number of electrons to have a full outer shell. Non-metals form negatively charged ions. When metals react with non-metals, the metal atoms lose electrons and the non-metal atoms gain electrons. This produces a compound with an **ionic structure**.

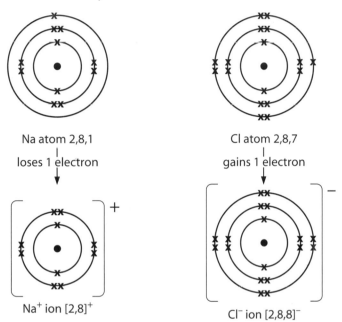

Na atom 2,8,1
loses 1 electron
Na⁺ ion [2,8]⁺

Cl atom 2,8,7
gains 1 electron
Cl⁻ ion [2,8,8]⁻

C Sodium atoms react with chlorine atoms to form the ionic compound sodium chloride.

The elements in Group 1 of the Periodic Table are called the **alkali metals**. They are all metals. When any Group 1 element reacts with a non-metal, an **ionic compound** is formed. The metal ion has a 1+ charge (for example, Li^+, Na^+, K^+, Rb^+, Cs^+) as the metal atom loses one electron.

The elements in Group 7 of the Periodic Table are called the **halogens**. They are all non-metals. When any Group 7 element reacts with a metal, an ionic compound is formed. The ion has a 1– charge (for example, F^-, Cl^-, Br^-, I^-, called halide ions) as the non-metal gains one electron.

9 When metals react with non-metals:
  a what happens to the metal atoms
  b what happens to the non-metal atoms
  c what type of substance is made?

5 What do all ions have in common in terms of their electronic structure?

6 Beryllium has an atomic number of 4. What is the electronic structure of a Be^{2+} ion?

7 Nitrogen has an atomic number of 7. What is the electronic structure of an N^{3-} ion?

8 Barium is in Group 2 of the Periodic Table. Predict the charge of barium ions.

D Sodium atoms react with chlorine atoms.

10 Potassium reacts with fluorine to form potassium fluoride.
  a Draw the electronic structure of the potassium and fluoride ions formed.
  b Explain why the ions formed are electrically charged.

Ionic substances

By the end of this topic you should be able to:

- describe the structure of ionic compounds
- describe what ionic bonding is
- describe and explain the properties of ionic compounds
- write the formulae of ionic compounds.

Many compounds are made up of ions. These substances are all solids at room temperature. There are billions of ions in the structure – from one edge of the solid right across to any of the others. The ions are packed together in an ordered, regular structure. It is a giant **lattice**.

Each ion is surrounded by ions of opposite charge. Each ion is attracted by a strong **electrostatic attraction** to all the ions of opposite charge surrounding it. This electrostatic attraction between positive and negative ions is known as **ionic bonding**.

1 What holds the ions together in an ionic compound?

2 What is ionic bonding?

3 Ionic compounds have a giant structure. What does this mean?

In order to melt and boil ionic compounds, the strong attractive forces between ions of opposite charges have to be overcome. This takes a lot of energy and so ionic compounds have high melting and boiling points.

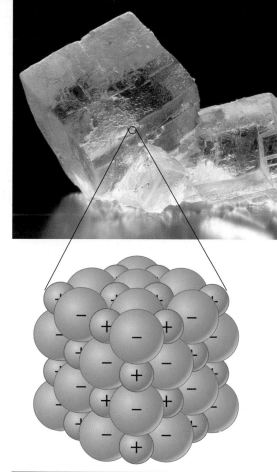

A The structure of sodium chloride (salt).

solid	liquid	gas

B The structure of an ionic compound as a solid, liquid and gas.

An electric current is a flow of charge. Ions are electrically charged particles. In solids, ions can only vibrate about fixed positions. Since the ions in solids do not flow they do not conduct electricity. In a liquid, ions are able to move around. When a solid melts it becomes a liquid. This allows the ions to flow and conduct electricity.

4 Why do ionic compounds have high melting and boiling points?

When you dissolve an ionic compound, the solution is a liquid. Ions in solution can flow and conduct electricity.

5 What is an electric current?

6 Explain why ionic compounds
   a do not conduct electricity as solids
   b can conduct electricity when melted
   c can conduct electricity when dissolved in water.

The formula of an ionic compound can be worked out using the ion charges. Table D includes some compound ions which are made up of several atoms joined together. For example, the nitrate ion (NO_3^-) contains one N and three Os and has an overall charge of 1−.

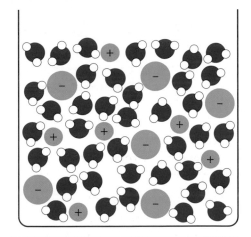

C Sodium chloride dissolved in water.

Positive Ions				Negative ions			
aluminium	Al^{3+}	lead	Pb^{2+}	bromide	Br^-	iodide	I^-
ammonium	NH_4^+	lithium	Li^+	carbonate	CO_3^{2-}	nitrate	NO_3^-
barium	Ba^{2+}	magnesium	Mg^{2+}	chloride	Cl^-	oxide	O^{2-}
calcium	Ca^{2+}	potassium	K^+	fluoride	F^-	sulfate	SO_4^{2-}
copper (II)	Cu^{2+}	silver	Ag^+	hydrogencarbonate	HCO_3^-	sulfide	S^{2-}
hydrogen	H^+	sodium	Na^+	hydroxide	OH^-		
iron (II)	Fe^{2+}	zinc	Zn^{2+}				
iron (III)	Fe^{3+}						

D Some positive and negative ions.

To work out the formula, the total charge of the positive ions must balance with the negative ions. In sodium oxide (Na_2O) there are two sodium (Na^+) ions (total charge 2+) for every one oxide (O^{2-}) ion (charge 2−). If a formula contains more than one of a compound ion, it must be put inside a bracket.

	Sodium oxide	Calcium sulfide	Aluminium oxide	Copper carbonate	Iron (III) nitrate	Calcium hydroxide	Ammonium sulfate
Positive ions	Na^+ Na^+	Ca^{2+}	Al^{3+} Al^{3+}	Cu^{2+}	Fe^{3+}	Ca^{2+}	NH_4^+ NH_4^+
Negative ions	O^{2-}	S^{2-}	O^{2-} O^{2-} O^{2-}	CO_3^{2-}	NO_3^- NO_3^- NO_3^-	OH^- OH^-	SO_4^{2-}
Formula	Na_2O	CaS	Al_2O_3	$CuCO_3$	$Fe(NO_3)_3$	$Ca(OH)_2$	$(NH_4)_2SO_4$

E Formulae of some ionic compounds.

The formula tells us the ratio of ions in the structure. The formula Al_2O_3 tells us that there are two aluminium (Al^{3+}) ions for every three oxide (O^{2-}) ions in aluminium oxide.

7 Use the table of ions to write the formula of:
   a potassium bromide
   b iron (III) oxide
   c aluminium hydroxide
   d calcium nitrate
   e zinc sulfate.

8 Calcium oxide has the formula CaO. Explain what this formula means.

9 Make a table with two columns to summarise the properties of ionic compounds. One column should give the properties and the other should explain the properties.

Simple molecular substances

By the end of this topic you should be able to:

- describe the structure of simple molecular substances
- describe and explain the properties of simple molecular substances
- represent the structure of covalent molecules.

1 What is a molecule?

2 What is a covalent bond?

Many substances are made up of **molecules**. Examples include oxygen (O_2) and water (H_2O). A molecule is a particle containing atoms joined by **covalent bonds**. A covalent bond is two shared electrons between two atoms.

Non-metal atoms need electrons to fill their outer energy level (shell). If non-metal atoms react with other non-metal atoms they can all fill their outer energy levels by sharing electrons. For example, chlorine atoms have seven electrons in their outer energy level and need one more to fill it. Hydrogen atoms have one electron in their outer energy level and need one more to fill it. The hydrogen and chlorine atoms can both fill their outer energy levels by sharing one electron from each atom.

To draw dot-cross diagrams for molecules:
- Draw a stick diagram (table B shows how many covalent bonds atoms form).
- Redraw the molecule without the sticks.
- Draw a dot and a cross instead of each stick (that is, ●X for a single bond, ●X●X for a double bond, etc.)
- Work out how many electrons there are in the outer energy level (shell) of each atom. Add in any of these electrons (in pairs) that are not already drawn in the covalent bonds.

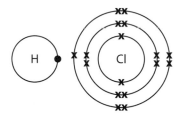 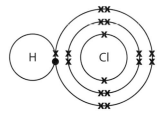 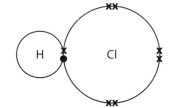

H atom	**Cl atom**	**HCl molecule**	**HCl molecule**	**HCl molecule**
		Formed by H and Cl atoms sharing two outer energy level (shell) electrons (one from each atom).	'Dot-cross' diagram showing outer energy level (shell) electrons only.	'Stick' diagram showing covalent bonds only. Each one is represented by one stick.

A Hydrogen chloride (HCl) molecule.

Atoms	Electrons in outer energy level	Electrons needed to fill outer energy level	Covalent bonds formed
H atoms	1	1	1
Group 7 atoms	7	1	1
Group 6 atoms	6	2	2
Group 5 atoms	5	3	3
Group 4 atoms	4	4	4

B Number of covalent bonds formed by different atoms.

3 a Draw a stick diagram for hydrogen fluoride (HF).
  b Draw a dot-cross diagram for hydrogen fluoride (HF).

	Molecule					
	H_2	Cl_2	H_2O	O_2	NH_3	CH_4
Stick diagram	H—H	Cl—Cl	H—O—H	O=O	H—N—H with N below	H—C—H with H above and below
Dot-cross diagram						

C Stick and dot-cross diagrams for common molecules.

Substances made up of molecules have a **simple molecular structure**. The covalent bonds between the *atoms* within a molecule are very strong. There are no bonds between the *molecules*. However, there are weak attractive forces between the molecules called **intermolecular forces**. They do not conduct electricity because the molecules are neutral.

H When simple molecular substances melt or boil, it is the weak forces between molecules that are overcome. The covalent bonds do not break. If the covalent bonds did break, then it would no longer be the same substance.

Simple molecular substances have low melting and boiling points because the forces between the molecules are weak. Some are solids at room temperature, but many are liquids or gases.

5 Why do simple molecular substances have low melting and boiling points?

The formula of a simple molecular substance tells us how many atoms of each type are in one molecule. For example, the formula CH_4 tells us that there is one carbon atom and four hydrogen atoms in one methane molecule.

4 Why do simple molecular substances not conduct electricity?

6 Bromine (Br_2) is a simple molecular substance.
  a Draw a stick diagram to represent bromine.
  b Draw a dot-cross diagram to represent bromine.
  c Predict what the physical properties of bromine are.
  d Explain what the formula of bromine (Br_2) means.
  H **e** Why does bromine have a low boiling point?

state	solid	liquid	gas
space-filling diagrams (better represents what molecules look like)			
stick diagrams			

D The structure of a simple molecular substance as a solid, liquid and gas.

Giant covalent substances

By the end of this topic you should be able to:

- describe the structure of giant covalent substances
- describe and explain the properties of giant covalent substances, including diamond, graphite and silicon dioxide.

In a molecule there are a specific number of *atoms* joined together by covalent bonds. There are no bonds between the *molecules*. As well as forming molecules, atoms that share electrons can also form **giant covalent structures**. These are sometimes called **macromolecular structures**.

In a giant covalent structure all the atoms are joined together by covalent bonds in a massive network. An example is diamond. In diamond, the carbon atoms are all linked together in a giant lattice from one edge of the diamond right across to the other. There are billions of carbon atoms joined together in one continuous structure by covalent bonds.

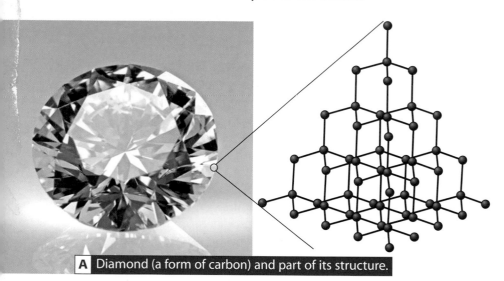

A Diamond (a form of carbon) and part of its structure.

B Graphite (a form of carbon).

Another example of a giant covalent substance is graphite. Diamond and graphite are both forms of the element carbon. Silicon dioxide is another example. Silicon dioxide has the formula SiO_2 which means that there is one silicon atom for every two oxygen atoms in the structure. In other words, there are twice as many oxygen atoms as silicon atoms.

In order to melt or boil a giant covalent substance, covalent bonds have to be broken. You need to provide energy (heat) to break bonds. Covalent bonds are very strong, and so you need to add a lot of heat to break them. This is why the melting point of diamond is over 3500 °C.

C Silicon dioxide (silica).

Solid	Liquid	Gas
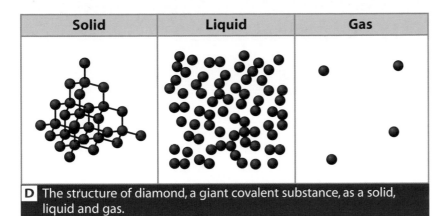		

D The structure of diamond, a giant covalent substance, as a solid, liquid and gas.

Although diamond and graphite are both forms of carbon and have giant covalent structures, they do not have exactly the same properties.

	Diamond	Graphite
Structure	Each C atom is joined to four others by covalent bonds.	Each C atom is joined to three others by covalent bonds. This forms layers. The layers are free to slide over each other. The layers are not bonded to each other.
Melting point	very high because lots of strong covalent bonds have to be broken	very high because lots of strong covalent bonds have to be broken
Hardness	very hard because the atoms are bonded in a rigid network	soft and slippery because the layers can slide over each other

E Comparison of diamond and graphite.

H

	Diamond	Graphite
Electrical and thermal conductivity	does not conduct heat or electricity, because there are no electrons free to move	conducts heat and electricity, because there is one electron from each carbon atom free to move (delocalised electrons) between the layers

F Comparison of electrical and thermal conductivity of diamond and graphite.

5 Explain why graphite conducts heat and electricity but diamond does not.

1 What type of bond joins the atoms together in a giant covalent substance?

2 Give three examples of giant covalent substances.

3 Why do giant covalent substances have very high melting points?

4 Explain why graphite is soft and slippery, but diamond is very hard.

6 a List as many similarities as you can between diamond and graphite.
  b List as many differences as you can between diamond and graphite.
  c Explain why graphite conducts electricity, but diamond does not.

H

Metallic substances

By the end of this topic you should be able to:

- describe and explain some of the properties of metallic substances
- describe the structure of metallic substances
- represent the bonding in metals using a diagram.

There are over 100 elements. Just over three-quarters of the elements are metals. A lot of use has been made of metals since the Bronze Age. Today we make much use of metals like iron, aluminium and copper. Metals all have the same type of structure, a **metallic** structure.

☐ non-metal

☐ metal

																		H	He
																		1	2
Li	**Be**													**B**	**C**	**N**	**O**	**F**	**Ne**
3	4													5	6	7	8	9	10
Na	**Mg**													**Al**	**Si**	**P**	**S**	**Cl**	**Ar**
11	12													13	14	15	16	17	18
K	**Ca**	**Sc**	**Ti**	**V**	**Cr**	**Mn**	**Fe**	**Co**	**Ni**	**Cu**	**Zn**	**Ga**	**Ge**	**As**	**Se**	**Br**	**Kr**
19	20	21	22	23	24	25	26	27	28	29	30	31	32	33	34	35	36
Rb	**Sr**	**Y**	**Zr**	**Nb**	**Mo**	**Tc**	**Ru**	**Rh**	**Pd**	**Ag**	**Cd**	**In**	**Sn**	**Sb**	**Te**	**I**	**Xe**
37	38	39	40	41	42	43	44	45	46	47	48	49	50	51	52	53	54
Cs	**Ba**	**La**	**Hf**	**Ta**	**W**	**Re**	**Os**	**Ir**	**Pt**	**Au**	**Hg**	**Tl**	**Pb**	**Bi**	**Po**	**At**	**Rn**
55	56	57	72	73	74	75	76	77	78	79	80	81	82	83	84	85	86
Fr	**Ra**	**Ac**	**Rf**	**Db**	**Sg**	**Bh**	**Hs**	**Mt**	**Ds**	**Rg**							
87	88	89	104	105	106	107	108	109	110	111							

A Metals in the Periodic Table.

As a solid, the metal atoms are packed close together in a regular structure. The outer energy level (shell) electron(s) are lost from each atom and become free to move around (delocalised) throughout the metal. This leaves a **giant structure** (lattice) of positive metal ions surrounded by delocalised electrons.

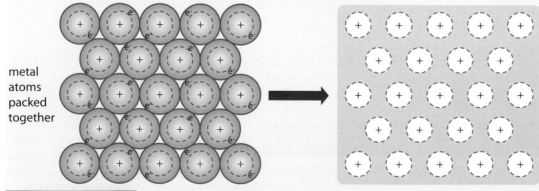

metal atoms packed together

Outer energy level (shell) electrons are lost from atoms leaving a lattice of positive metal ions surrounded by delocalised electrons. The shaded area represents the delocalised electrons.

B Metallic bonding.

The structure is held together by the electrostatic attraction between the positive metal ions and the negative delocalised electrons. This attraction is strong and metals have high melting and boiling points.

1 a Describe the structure of a metal.
  b Describe the bonding in a metal.

2 Why do metals have high melting points?

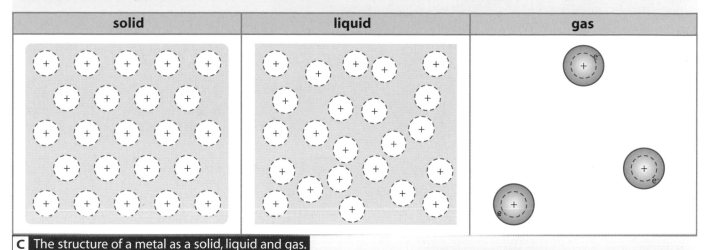

solid	liquid	gas

C The structure of a metal as a solid, liquid and gas.

Pans are often made from metals because they conduct heat well to cook the food. Electrical wires are made of metals because they conduct electricity. Metals conduct heat and electricity because the delocalised electrons are free to move.

D Metals conduct heat.

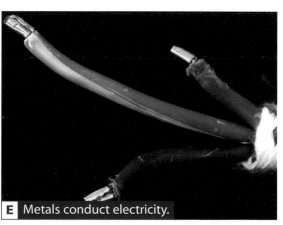

E Metals conduct electricity.

3 Explain why metals can be used to make:
  a pans b electrical wires.

4 Explain why metals can be bent and shaped.

Metals are strong. They can be bent and hammered into shape. This is because the layers of metal ions can slide over each other while keeping the strong attraction between the positive metal ions and the delocalised electrons.

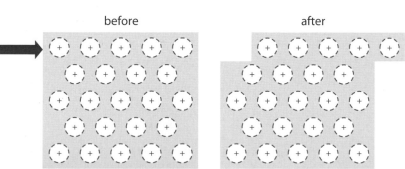

before   after

F Metals can be bent and shaped.

5 Copper is used to make electrical wires.
  a Explain why copper is used to make electrical wires.
  b Draw a diagram to show the structure of copper.
  c Use your diagram to explain why copper conducts electricity.
  d Use your diagram to explain why copper can be stretched into wires.

Types of structure

By the end of this topic you should be able to:

- describe the three types of bonding
- describe the five types of structure
- relate the properties of each type of structure to its use.

There are millions of different substances but just three different types of bonding and five different types of structure. The three different types of bonding are shown in table A.

1 Name the three types of bonding.

2 Describe what the following are:
   a a covalent bond
   b ionic bonding
   c metallic bonding.

Type of bonding	Description	Types of structure that have this type of bonding	Examples
Covalent	two shared electrons between two atoms	simple molecular giant covalent	water diamond
Ionic	the attraction between positive ions and negative ions	ionic	salt, copper sulfate
Metallic	the attraction between positive metal ions and delocalised outer shell electrons	metallic	iron, aluminium

A The three types of bonding.

B Hydrogen and chlorine have simple molecular structures.

C Sodium chloride and sodium hydroxide have ionic structures.

Table E summarises the five different types of structure. The summary includes **monatomic** structures which have not been described up to now. The elements in Group 0 of the Periodic Table, the noble gases, are the only six substances with a monatomic structure. There is no bonding at all in these substances. They exist as separate atoms.

3 Which type of structure:
   a has high melting and boiling points, and conducts electricity as a solid
   b has high melting and boiling points, and conducts electricity when melted but not as a solid
   c has low melting and boiling points, and never conducts electricity
   d has very high melting and boiling points, and does not conduct electricity
   e has very low melting and boiling points, and never conducts electricity?

4 Decide which type of structure the six substances in table D have.

Substance	Melting point (°C)	Boiling point (°C)	Electrical conductivity as...	
			...solid	...liquid
A	750	1392	✗	✓
B	−269	−268	✗	✗
C	6	80	✗	✗
D	327	1744	✓	✓
E	−23	77	✗	✗
F	1703	2230	✗	✗

D

5 Draw a flow diagram to show how the five different types of structure could be distinguished by simple tests.

	Monatomic	Simple molecular	Giant covalent	Ionic	Metallic
Description of the structure	Made up of individual atoms that are not joined together. **H** There are only very weak forces between the atoms.	Made up of many individual molecules. The molecules are not joined to each other. Within the molecules the atoms are joined by strong covalent bonds. **H** There are only very weak forces between the molecules.	A **giant structure**. The atoms are all joined together by strong covalent bonds in a continuous network.	A giant structure of positive and negative ions. There is a strong attraction between the positive and negative ions.	**H** A giant structure of positive metal ions surrounded by delocalised outer energy level (shell) electrons. There is a strong attraction between the positive metal ions and the delocalised electrons.
Type of bonding	none	covalent	covalent	ionic	metallic
Which substances have this structure	Group 0 elements	some non–metal elements and compounds made from non-metals	diamond, graphite and silicon dioxide (SiO_2)	compounds made from metals and non-metals	metals
Solid					**H**
Liquid					**H**
Gas					**H**
Formula	The symbol from the Periodic Table (for example, Ar).	Gives the number of atoms of each type in one molecule. For example CH_4 means there is 1 C atom and 4 H atoms in each molecule.	Gives the ratio of atoms of each type in the structure. For example SiO_2 means there are twice as many O atoms in the lattice as Si atoms.	Gives the ratio of ions of each type in the structure. For example $MgCl_2$ means there are twice as many Cl^- ions in the lattice as there are Mg^{2+} ions.	The symbol from the Periodic Table (for example, Mg).
Melting and boiling points	very low **H** due to very weak forces between the atoms	low **H** due to weak forces between the molecules	very high due to many strong covalent bonds which have to be broken	high due to the strong attraction between the positive and negative ions	high **H** due to the strong attraction between the positive metal ions and the delocalised electrons
Do they conduct heat and electricity?	no The atoms are neutral.	no The molecules are neutral.	no (for example, diamond, silicon dioxide) yes (graphite) **H** Diamond and silicon dioxide have no delocalised electrons; graphite has delocalised electrons.	no (solid) yes (liquid/solution) Ions cannot move as a solid, but can when a liquid or in solution.	yes **H** Outer energy level (shell) electrons are free to move (delocalised).
Are they soluble in water?	insoluble	sometimes soluble	insoluble	usually soluble	insoluble

E The five different types of structure.

Smart materials and nanoscience

By the end of this topic you should be able to:

- describe what smart materials and nanoparticles are
- give some examples of smart materials and nanoparticles
- evaluate developments and applications of smart materials and nanoscience.

Smart materials are materials which have one or more properties that change in different conditions. Their properties may change with temperature, light levels, pH, electric fields or magnetic fields.

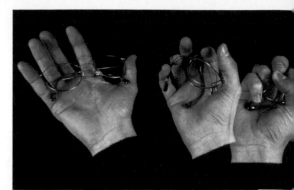

A These spectacles are made of shape-memory alloy and return to their original shape.

Uses of smart materials

Shape-memory alloys and shape-memory polymers are examples of smart materials. Some change shape as their temperature changes. Shape-memory polymers are used in heat shrink-wrapping and tubes for electrical cables. Shape-memory alloys are used in electrical devices such as fire-alarm systems that trigger sprinklers.

Some smart materials change colour with changes in temperature or light. For example, some kettles change colour as the water gets hotter. There are also light-sensitive lenses for spectacles that darken in bright light.

B Some smart materials change colour as the temperature changes.

1 What are smart materials?

2 Give five examples of things that can change the properties of smart materials.

3 **a** Give two uses of smart materials.
   b Explain how the smart materials work in the uses you have given in part a.

Most atoms are between 0.1 and 0.2 nm (nanometres) in size. **Nanoscience** is the study of structures that are between 1 to 100 nm in size. These structures are called **nanoparticles**. One nanometre is one millionth of a millimetre. Nanoparticles contain a few hundred atoms. **Nanomaterials** are made from nanoparticles.

Nanoparticles can have very different properties to the same materials in bulk. They can be stronger than the material in bulk form, or they may conduct heat or electricity in a different way. Nanoparticles of gold appear red rather than the yellow colour of the bulk material. They can also act as a catalyst in reactions for which bulk gold cannot. Nanoparticles of silver kill bacteria, but the bulk material does not.

One of the reasons why nanoparticles have different properties to the bulk material is that the smaller nanoparticles have a much larger surface area to volume ratio than the bulk material. This means that a much higher fraction of the atoms are on the surface.

Uses of nanoscience

Scientists are using the unusual properties of nanoparticles to create new technologies. Some uses are:
- to make computers more powerful and faster
- to kill cancer cells
- to make construction materials stronger and lighter
- as new catalysts (for example, using gold as a catalyst in oxidation reactions)
- as new coatings (for example, on glass to make self-cleaning windows that repel water and break down dirt)
- as highly selective sensors (for example, to detect certain chemicals).

There are some concerns though about the effects of nanoparticles on people. It is possible that some nanomaterials are toxic, even if the bulk material is not. Further research needs to be done to find out what the effects might be.

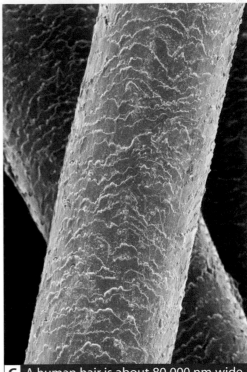

C A human hair is about 80 000 nm wide, many times wider than nanoparticles.

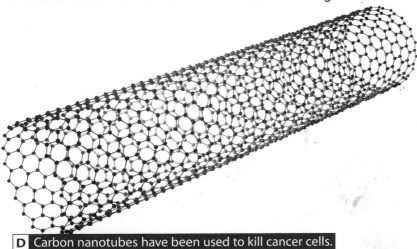

D Carbon nanotubes have been used to kill cancer cells.

4 What is a nanoparticle?

5 Why are nanoparticles useful?

6 Why do nanoparticles often have different properties to the bulk material?

7 Give three uses of nanoparticles.

8 Give one possible problem with using nanoparticles.

9 Smart materials and nanoparticles are new areas of science. Write an entry for an online encyclopaedia for the following words:
   a nanoparticles
   b nanoscience
   c smart materials
   d shape memory materials.
   Use examples to illustrate your descriptions.

Relative masses

By the end of this topic you should be able to:

- explain what relative atomic and relative formula masses are
- calculate the relative formula mass of a substance
- calculate the percentage by mass of an element in a compound.

Sometimes it can be useful to measure the mass of things using **relative mass** instead of the mass in grams or kilograms. For example, the masses in table A are all relative to the mass of a man. The elephant has a relative mass of 60 which means it is 60 times heavier than the man.

	Man	Lion	Car	Elephant
Mass	75 kg	150 kg	1650 kg	4500 kg
Mass relative to man	$\frac{75}{75} = 1$	$\frac{150}{75} = 2$	$\frac{1650}{75} = 22$	$\frac{4500}{75} = 60$

A Some masses relative to an adult human.

The relative mass of an individual atom is the same as its mass number. The mass number is the number of protons plus the number of neutrons in an atom.

H Atoms are tiny and very light. For example, ^{12}C atoms have a mass of 0.000 000 000 000 000 000 000 020 g. Working with numbers like this is difficult so we use relative masses. This scale compares the mass of an atom to one-twelfth the mass of a ^{12}C atom. This means that ^{12}C atoms have a relative mass of 12. ^1H atoms have a relative mass of 1 which means they are 12 times lighter than ^{12}C atoms. ^{24}Mg atoms have a relative mass of 24 which means they are twice as heavy as ^{12}C atoms.

1 Use the Periodic Table on page 225 to find the relative atomic mass of:
- **a** nitrogen
- **b** oxygen
- **c** silicon
- **d** copper.

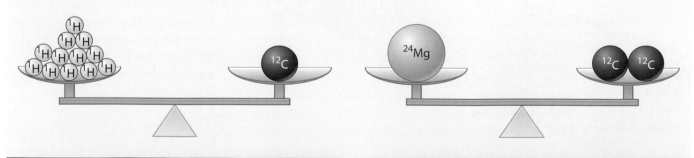

B Twelve ^1H atoms have the same mass as one atom of ^{12}C.　　**C** One ^{24}Mg atom has the same mass as two atoms of ^{12}C.

The atoms of most elements are a mixture of different isotopes. The **relative atomic mass (A_r)** of an element is the average relative mass of these different atoms. For example, about three-quarters of chlorine atoms are ^{35}Cl atoms and about one-quarter are ^{37}Cl atoms. The average relative mass of these atoms is 35.5, so the relative atomic mass of chlorine is 35.5.

The **relative formula mass (M_r)** of a substance is the sum of the relative atomic masses of all the atoms in the formula. For example, H_2O contains two hydrogen atoms and one oxygen atom, and so has a relative formula mass of 18.

For example:
Calculate the M_r of aluminium chloride, $AlCl_3$. (Relative atomic masses: Al = 27, Cl = 35.5)
$$M_r = 27 + (3 \times 35.5) = 133.5$$

Calculate the M_r of calcium nitrate, $Ca(NO_3)_2$. (Relative atomic masses: Ca = 40, N = 14, O = 16)
$$M_r = 40 + (2 \times 14) + (6 \times 16) = 164$$

sodium chloride (NaCl) $M_r = 58.5$	hydrogen (H_2) $M_r = 2$	chlorine (Cl_2) $M_r = 71$

sodium hydroxide (NaOH) $M_r = 40$

E The M_r of substances involved in the electrolysis of salt water.

Using relative masses it is possible to calculate what percentage of the mass of a substance is a particular element. This can be done using the following equation:

$$\text{\% by mass of an element in a compound} = 100 \times \frac{\text{relative mass of all the atoms of that element}}{M_r}$$

For example, water molecules have a relative mass of 18, of which 16 is the mass of the oxygen atom and 2 is the total mass of the two hydrogen atoms:

$$\text{\% O in } H_2O = 100 \times \frac{16}{18} = 88.9\%$$

Similarly,

$$\text{\% O in } Ca(NO_3)_2 = 100 \times \frac{(6 \times 16)}{164} = 58.5\%$$

2 a What is the difference between mass number and relative atomic mass (A_r)?
b What is the mass number of the commonest chlorine atoms?
c What is the relative atomic mass of chlorine?

Calculate the M_r of water, H_2O.
(relative atomic masses: H = 1, O = 16)

$$M_r = (2 \times 1) + 16 = 18$$

D The relative mass of a water molecule is 18.

3 Calculate the relative formula mass of the following substances (use the Periodic Table on page 225 to find relative atomic masses):
a methane (CH_4)
b oxygen (O_2)
c magnesium chloride ($MgCl_2$)
d aluminium sulfate ($Al_2(SO_4)_3$).

4 Calculate the percentage by mass of:
a C in propane (C_3H_8)
b Cl in sodium chloride (NaCl)
c N in ammonium nitrate (NH_4NO_3)
d O in aluminium sulphate ($Al_2(SO_4)_3$).

5 The *percentage by mass* of hydrogen in methane is 25%. This can be worked out using the *relative atomic masses* of carbon and hydrogen and the M_r of methane. Explain what is meant by the terms in italics and show how the 25% is worked out.

Moles

By the end of this topic you should be able to:

P

H
- define what a mole is
- use the formula: mass = M_r × moles

A pair is two of anything. A dozen is 12 of anything. A **mole** is 602 204 500 000 000 000 000 000 of anything. Scientists count particles using moles. A mole of particles of many substances fits well into a boiling tube or beaker. For example, a mole of water molecules has a mass of 18 g. A small glass of water contains about 10 moles of water molecules.

The number 602 204 500 000 000 000 000 000 was carefully chosen so that the mass of that number of particles equals the relative formula mass (M_r) in grams. For example, the M_r of water is 18, and so the mass of one mole of water molecules is 18 g. The M_r of carbon dioxide is 44, so the mass of one mole of carbon dioxide molecules is 44 g.

A There are about 10 moles of water molecules in this glass.

H You need to be able to work out how many moles of a substance you have from the mass. If one mole of water molecules has the mass 18 g, then 2 moles has the mass 36 g (2 × 18). The general equation is shown below. This can be remembered by thinking of 'Mr Moles'! It can also be shown with this triangle. By covering the quantity you need to calculate you can find the equation to use.

mass (g) = M_r × moles

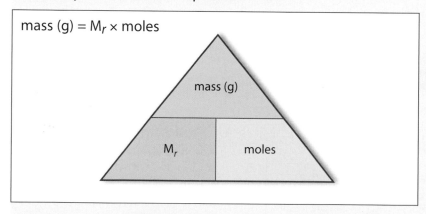

1 mg	= 0.001 g
1 kg	= 1 000 g
1 tonne	= 1 000 000 g

B Mass units.

P

1 What is the equation you would use to work out the number of moles of a substance using mass and M_r?

2 What is the equation you would use to work out the M_r of a substance using mass and number of moles?

3 Write the following masses in grams:
   a 20 kg
   b 5 tonnes
   c 50 mg.

P

Substance	Formula	M_r	Mass	Moles
Water	H_2O	18	36 g	$= \dfrac{mass\ (g)}{M_r}$ $= \dfrac{36}{18} = 2$
Calcium carbonate	$CaCO_3$	100	1 tonne	$= \dfrac{mass\ (g)}{M_r}$ $= \dfrac{1\ 000\ 000}{100} = 10\ 000$
Carbon dioxide	CO_2	44	$= M_r \times moles$ $= 44 \times 0.25$ $= 11\ g$	0.25
Ammonium nitrate	NH_4NO_3	80	$= M_r \times moles$ $= 80 \times 150$ $= 12\ 000\ g$	150
Unknown gas	unknown	$= \dfrac{mass\ (g)}{moles}$ $= \dfrac{3.2}{0.10} = 32$	3.2 g	0.10
Unknown solid	unknown	$= \dfrac{mass\ (g)}{moles}$ $= \dfrac{0.61}{0.005} = 122$	0.61 g	0.005

C Example calculations using the equation mass (g) = M_r × moles.

4 How many moles are there in the following:
  a 8 g of SO_3
  b 1 kg of Fe_2O_3?

5 What mass would the following quantities have:
  a 2.5 moles of Cl_2
  b 0.01 moles of Na_2CO_3?

6 What is the M_r of the following substances:
  a 0.05 moles of cyclohexane, which has mass 4.2 g
  b 0.001 moles of aspirin, which has mass 0.18 g?

D One mole of salt.

7 The mass of 10 moles of oxygen gas (O_2) is 320 g. Explain what is meant by the word *moles* and show how this mass can be calculated.

Reacting-mass calculations

By the end of this topic you should be able to:

- calculate the mass of substances involved in chemical reactions from balanced equations.

The equation in A shows the reaction between hydrogen and nitrogen to make ammonia. You can use the equation to see how many particles of each substance are involved in the reaction.

nitrogen	+	hydrogen	⟶	ammonia
N_2	+	$3H_2$	⟶	$2NH_3$

1 N_2 molecule	3 H_2 molecules	2 NH_3 molecules
12 N_2 molecules	36 H_2 molecules	24 NH_3 molecules
1 dozen N_2 molecules	3 dozen H_2 molecules	2 dozen NH_3 molecules
602 204 500 000 000 000 000 000 N_2 molecules	1 806 613 500 000 000 000 000 000 H_2 molecules	1 204 409 000 000 000 000 000 000 NH_3 molecules
1 mole of N_2 molecules	3 moles of H_2 molecules	2 moles of NH_3 molecules

A Hydrogen and nitrogen react together to form ammonia.

You can use this to calculate the masses of chemicals that react with each other and the mass of chemicals produced. The rules for doing this are:

- calculate the number of moles of the substance whose mass is given (moles = mass (g) ÷ M_r)
- use the chemical equation to work out how many moles of the substance asked about are used or made
- calculate the mass of the substance asked for (mass (g) = M_r × moles).

Using reactions

Example 1

Iron is made when aluminium reacts with iron oxide. This reaction is used to weld railway lines together. What mass of aluminium is needed to react with 640 g of iron oxide? (Relative atomic masses: O = 16, Al = 27, Fe = 56)

$$Fe_2O_3 + 2Al \longrightarrow 2Fe + Al_2O_3$$

$$\text{moles of } Fe_2O_3 = \frac{\text{mass (g)}}{M_r} = \frac{640}{160} = 4$$

moles of Al = moles of Fe_2O_3 × 2 = 4 × 2 = 8
mass of Al = M_r × moles = 27 × 8 = $\boxed{216 \text{ g}}$

B Welding railway lines together.

Example 2

Calcium hydroxide (slaked lime) is used by farmers to neutralise acidic soil. Calcium hydroxide is made by adding water to calcium oxide (quicklime). What mass of calcium hydroxide is made from 14 kg of calcium oxide? (Relative atomic masses: H = 1, O = 16, Ca = 40)

$$CaO + H_2O \rightarrow Ca(OH)_2$$

$$\text{moles of CaO} = \frac{\text{mass (g)}}{M_r} = \frac{14\,000}{56} = 250$$

moles of $Ca(OH)_2$ = moles of CaO = 250

mass of $Ca(OH)_2$ = M_r × moles = 74 x 250 = $\boxed{18\,500 \text{ g}}$

Example 3

Propane is often used as the fuel in gas fires and barbecues. What mass of oxygen is needed to burn 110 g of propane? (Relative atomic masses: H = 1, C = 12, O = 16)

$$C_3H_8 + 5O_2 \rightarrow 3CO_2 + 4H_2O$$

$$\text{moles of } C_3H_8 = \frac{\text{mass (g)}}{M_r} = \frac{110}{44} = 2.5$$

moles of O_2 = moles of C_3H_8 × 5 = 2.5 × 5 = 12.5

mass of O_2 = M_r × moles = 32 × 12.5 = $\boxed{400 \text{ g}}$

Example 4

Titanium is a metal. One of its uses is to make replacement hip joints. Titanium can be made by reacting titanium chloride with sodium. What mass of titanium chloride reacts with 460 g of sodium? (Relative atomic masses: Na = 23, Cl = 35.5, Ti = 48)

$$TiCl_4 + 4Na \rightarrow Ti + 4NaCl$$

$$\text{moles of Na} = \frac{\text{mass (g)}}{M_r} = \frac{460}{23} = 20$$

moles of $TiCl_4$ = moles of Na ÷ 4 = 20 ÷ 4 = 5

mass of $TiCl_4$ = M_r × moles = 190 × 5 = $\boxed{950 \text{ g}}$

1 What mass of hydrogen is produced when 96 g of magnesium reacts with hydrochloric acid? (Relative atomic masses: H = 1, Mg = 24)
$$Mg + 2HCl \rightarrow MgCl_2 + H_2$$

2 What mass of oxygen reacts with 46 g of sodium? (Relative atomic masses: O = 16, Na = 23)
$$4Na + O_2 \rightarrow 2Na_2O$$

3 What mass of water is formed when 1 kg of methane burns? (Relative atomic masses: H = 1, C = 12, O = 16)
$$CH_4 + 2O_2 \rightarrow CO_2 + 2H_2O$$

4 What mass of aluminium reacts with 10.65 g of chlorine to make aluminium chloride? (Relative atomic masses: Al = 27, Cl = 35.5)
$$2Al + 3Cl_2 \rightarrow 2AlCl_3$$

5 Draw a flow diagram to show how to do reacting-mass calculations.

C Reacting-mass calculations in the electrolysis of salt water.

HYDROGEN

Empirical and molecular formulae

H

By the end of this topic you should be able to:

- explain what an empirical formula represents
- explain what a molecular formula represents
- work out empirical and molecular formulae.

All substances have an **empirical formula**. This represents the simplest whole number ratio of atoms (or ions) of each element in a substance. For example, the empirical formula of silicon dioxide (silica) is SiO_2. This means that the ratio of silicon (Si) atoms to oxygen (O) atoms is 1:2. This means there are twice as many oxygen atoms as there are silicon atoms.

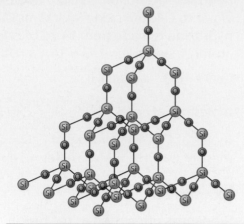

A Silicon dioxide (empirical formula = SiO_2).

Substances that are made of molecules also have a **molecular formula**. This represents the number of atoms of each element in one molecule. For example, ethene has the molecular formula C_2H_4, which means that there are two carbon (C) and four hydrogen (H) atoms in one molecule. It has the empirical formula CH_2, meaning that the simplest ratio of carbon to hydrogen atoms is 1:2.

To summarise:

- empirical formula gives the simplest ratio of atoms of each element in a substance
- molecular formula gives the number of atoms of each element in one molecule.

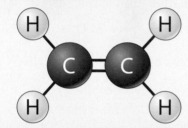

B Ethene (molecular formula = C_2H_4, empirical formula = CH_2).

For some molecules, the molecular formula is the same as the empirical formula. For example, the molecular formula of water is H_2O, meaning that there are two hydrogen (H) atoms and one oxygen (O) atom in each molecule. The empirical formula is also H_2O, meaning that the simplest ratio of hydrogen to oxygen atoms is 2:1.

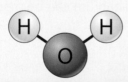

C Water (molecular formula = H_2O, empirical formula = H_2O).

Substance	Empirical formula	Molecular formula
Ethene	CH_2	C_2H_4
Propene	CH_2	C_3H_6
Water	H_2O	H_2O
Silicon dioxide (SiO_2)	SiO_2	no molecular formula (not made of molecules)
Sodium chloride	NaCl	no molecular formula (not made of molecules)
Ethane	CH_3	C_2H_6
Propane	C_3H_8	C_3H_8
Glucose	CH_2O	$C_6H_{12}O_6$

D Empirical and molecular formulae.

1 a Propene has the molecular formula C_3H_6. Explain what this means.
   b Propene has the empirical formula CH_2. Explain what this means.

2 The molecular formulae of some substances are shown. What is the empirical formula of each substance:
   a benzene = C_6H_6 b butane = C_4H_{10} c pentane = C_5H_{12}?

Substances can be analysed to find what elements they are made from. This analysis gives the mass (or percentage by mass) of each element in the substance. You can then calculate the empirical formula from this.

E Salt has the empirical formula NaCl.

Method	Example 1			Example 2	
	Analysis of a compound found that it contained 2.4 g of carbon, 0.4 g of hydrogen and 3.2 g of oxygen. (A_r: C = 12, H = 1, O = 16)			Analysis of a compound found that it contained 70% iron and 30% oxygen by mass. (A_r: Fe = 56, O = 16)	
1 Make a column for each element.	C	H	O	Fe	O
2 Divide the mass (or percentage) of each element by its relative atomic mass (A_r).	$\frac{2.4}{12} = 0.2$	$\frac{0.4}{1} = 0.4$	$\frac{3.2}{16} = 0.2$	$\frac{70}{56} = 1.25$	$\frac{30}{16} = 1.875$
3 Simplify this ratio by dividing all the answers by the smallest.	$\frac{0.2}{0.2} = 1$	$\frac{0.4}{0.2} = 2$	$\frac{0.2}{0.2} = 1$	$\frac{1.25}{1.25} = 1$	$\frac{1.875}{1.25} = 1.5$
4 Find the simplest whole number ratio. The numbers are from real experiments so may not be exact whole numbers. However, you may need to multiply the answers by a number such as 2, 3 or 4 to get close to whole numbers.	1	2	1	$1 \times 2 = 2$	$1.5 \times 2 = 3$
5 Write the empirical formula.	CH_2O			Fe_2O_3	

F How to calculate the empirical formula.

The molecular formula of a substance made of molecules can be found from the empirical formula and the relative formula mass (M_r):

$$\text{molecular formula} = \text{empirical formula} \times \frac{M_r}{M_r \text{ of empirical formula}}$$

Example 1

Find the molecular formula of a compound with empirical formula CH_2 and relative formula mass (M_r) = 56.

$$\text{molecular formula} = CH_2 \times \frac{56}{14} = CH_2 \times 4 = C_4H_8$$

Example 2

Find the molecular formula of a compound with empirical formula NH_3 and relative formula mass (M_r) = 17.

$$\text{molecular formula} = NH_3 \times \frac{17}{17} = NH_3 \times 1 = NH_3$$

3 Find the empirical formulae of the following compounds from the data given:
   a sulfur 40%, oxygen 60% (relative atomic masses: S = 32, O = 16)
   b potassium 18.2%, iodine 59.4%, oxygen 22.4% (relative atomic masses: K = 39, I = 127, O = 16).

4 Find the molecular formula of a compound with the empirical formula CH_2 and M_r = 70. (relative atomic masses: C = 12, H = 1)

5 Butane has the molecular formula C_4H_{10} and the empirical formula C_2H_5. Explain what these formulae mean and draw a flow diagram to show how to work out an empirical formula from experimental data.

Yields and atom economy

By the end of this topic you should be able to:

H • balance equations
• explain why less product is made than calculated

H • calculate the percentage yield for a reaction
• explain what atom economy is and why reactions with a high atom economy are important for sustainable development and economic reasons

H • be able to calculate the atom economy for a reaction.

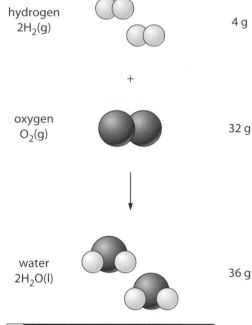

hydrogen $2H_2(g)$		4 g
+		
oxygen $O_2(g)$		32 g
water $2H_2O(l)$		36 g

A Chemical reaction to form water.

Chemical equations show how many particles of substances are involved in chemical reactions. For example, in the equation in diagram A, two molecules of hydrogen (H_2) react with one molecule of oxygen (O_2) to make two molecules of water (H_2O). The symbols after each formula tell us what state each substance is in.

When you make a new substance by a chemical reaction, you may not get all the expected amount of product. For example, if you react 4 g of hydrogen with 32 g of oxygen, you may get less than 36 g of water. There are a number of reasons for this, including:
• the reaction may be reversible (both the forwards and backwards reaction can take place)
• some of the product may be lost when it is separated from the reaction mixture
• some of the reactants may react in other reactions.

(s)	solid
(l)	liquid
(g)	gas
(aq)	aqueous (dissolved in water)

H The amount of a product obtained is known as the **yield**. The **percentage yield** compares the expected amount of product to the amount obtained.

$$\% \text{ yield} = \frac{\text{mass of product obtained}}{\text{maximum theoretical mass of product}} \times 100$$

For example, if you only obtained 27 g of water from the above reaction where the maximum theoretical yield is 36 g, then the percentage yield is 75%. This means that only 75% of the water that could have been produced has been obtained.

$$\% \text{ yield} = \frac{27}{36} \times 100 = 75\%$$

Chemists try to improve their methods and techniques to get as high a percentage yield as possible.

1 Calculate the percentage yields shown in table B.

	Theoretical maximum mass of product	Mass of product obtained
a	10 g	4 g
b	2 g	1.6 g
c	50 g	39 g

B

Industrial processes and atom economy

Chemists also try and achieve a high **atom economy (atom utilisation)** in reactions. The atom economy measures the amount of the starting materials that end up as useful products. In a reaction with a high atom economy, most of the mass of the starting materials is found in the products.

Sustainable development is doing what we need to meet peoples' needs and improve their lives today in a way that does not stop people from meeting needs in the future. Often, the higher the atom economy, the better a process is for sustainable development because there is less waste. However, the waste in many industrial processes is used in other processes.

H

$$\text{atom economy} = \frac{\text{mass of wanted product from equation}}{\text{total mass of products from equation}} \times 100$$

For example, ethanol is made in two different ways. It can be made by fermentation of sugars or by the reaction of ethene with steam.

Fermentation of sugars

	glucose	ethanol	carbon dioxide
$C_6H_{12}O_6(aq) \rightarrow 2CH_3CH_2OH(aq) + 2CO_2$
180 g		92 g	88 g

180 g products

Hydration of ethane

ethene   steam   ethanol

$CH_2=CH_2(g) + H_2O(g) \rightarrow CH_3CH_2OH(g)$

28 g   18 g   46 g

2 a What is sustainable development?
  b Why is a high atom economy good for sustainable development?

C Ethanol can be made from sugar cane.

$$\text{atom economy} = \frac{92}{180} \times 100 = 51\%$$

Only 92 g of the 180 g of products is ethanol. This means that 51% of the mass of the products is ethanol, while the other 49% is waste.

$$\text{atom economy} = \frac{46}{46} \times 100 = 100\%$$

All of the 46 g of products is ethanol.

3 Calculate the atom economy shown in table D.

	Mass of products in equation	Mass of wanted product in equation
a	80 g	64 g
b	128 g	96 g
c	140 g	122 g

D

4 a In the production of aspirin, less aspirin is made than should be produced by calculation. Give some reasons why.
H **b** The production of aspirin has a 90% yield and 75% atom economy. Explain what these figures mean.

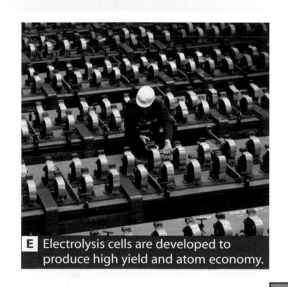

E Electrolysis cells are developed to produce high yield and atom economy.

Electrolysis

By the end of this topic you should be able to:

- describe what happens in electrolysis
- represent electrolysis by half equations
- define oxidation and reduction in terms of the transfer of electrons.

Ionic substances are compounds made up of positive and negative ions. Ionic substances are all solids at room temperature. As solids, they do not conduct electricity because the ions cannot move around.

When ionic substances are melted or dissolved in water they can conduct electricity. This is because the ions are free to move about within the liquid or solution. When they conduct electricity, the ionic substances break down (decompose), often into the elements they are made from. This process is called electrolysis.

Opposite charges attract. During electrolysis, the negative ions are attracted towards the positive electrode where they lose electrons (**oxidation**). These electrons move around the circuit towards the negative electrode. The positive ions are attracted towards the negative electrode where they gain electrons (**reduction**).

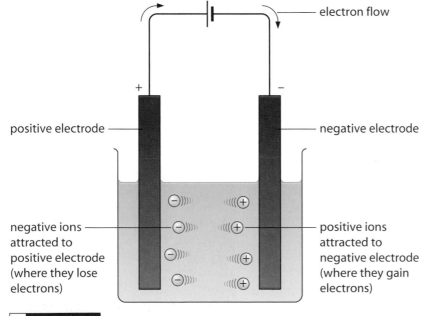

A Electrolysis.

You have met the terms 'oxidation' and 'reduction' before. The words can be used to describe reactions involving the gain or loss of oxygen. However, the same words can be used to describe reactions involving the transfer of electrons.

1 Which type of substance is broken down by electrolysis?

2 Why do ionic substances not conduct electricity as solids?

3 Why do ionic substances conduct electricity when they are melted or dissolved in water?

4 a Which electrode are the negative ions attracted to?
  b Why are they attracted to it?
  c What happens to the ions?
  d What happens to the electrons that are lost?

5 a Which electrode are the positive ions attracted to?
  b Why are they attracted to it?
  c What happens to the ions?

	In terms of oxygen	In terms of electrons
Oxidation	gain of oxygen	loss of electrons
Reduction	loss of oxygen	gain of electrons

B Oxidation in terms of oxygen and electrons.

Oxidation
Is
Loss

Reduction
Is
Gain

} of electrons

H You can write equations for the reactions that take place at the two electrodes. Each equation is called a **half equation**. The half equations show how many electrons are lost or gained. These equations should be balanced. Some of the elements formed are made of molecules containing two atoms (for example, H_2, O_2, Cl_2, Br_2, I_2). When their atoms form at the electrode, they pair up to make these molecules.

C The definition of oxidation and reduction in terms of electrons can be remembered using 'OIL RIG'.

Molten substance	Formula	Ions contained	Negative electrode		Positive electrode	
			Equation	**Product**	**Equation**	**Product**
Sodium chloride	$NaCl(l)$	Na^+, Cl^-	$Na^+ + e^- \rightarrow Na$	sodium	$2Cl^- \rightarrow Cl_2 + 2e^-$	chlorine
Lead bromide	$PbBr_2(l)$	Pb^{2+}, Br^-	$Pb^{2+} + 2e^- \rightarrow Pb$	lead	$2Br^- \rightarrow Br_2 + 2e^-$	bromine
Aluminium oxide	$Al_2O_3(l)$	Al^{3+}, O^{2-}	$Al^{3+} + 3e^- \rightarrow Al$	aluminium	$2O^{2-} \rightarrow O_2 + 4e^-$	oxygen

D Half equations in the electrolysis of some ionic substances.

6 For each of these half equations:
a balance the equation
   (i) $Mg^{2+} + \underline{\quad} e^- \rightarrow Mg$
   (ii) $\underline{\quad} I^- \rightarrow I_2 + \underline{\quad} e^-$
   (iii) $Cu^{2+} + \underline{\quad} e^- \rightarrow Cu$
   (iv) $\underline{\quad} H^+ + \underline{\quad} e^- \rightarrow H_2$
b state which electrode each half equation happens at
c state whether each half equation involves oxidation or reduction.

Using electrolysis

Electrolysis has many uses. One large-scale use is to extract reactive metals from their ores, such as aluminium.

E Aluminium is extracted from bauxite by electrolysis.

7 Potassium iodide is an ionic substance with the formula KI. It contains potassium (K^+) and iodide (I^-) ions. In the electrolysis of molten potassium iodide, for *each* electrode:
a predict what would be formed
b state whether the reaction is oxidation or reduction
c write a half equation to represent the process.

Electrolysis of solutions

By the end of this topic you should be able to:

• predict the products of electrolysis of solutions.

Ionic substances can also conduct electricity when dissolved in water because the ions can move around. However, the electrolysis of solutions is more complicated than the electrolysis of molten ionic compounds. This is because in water there are some hydrogen (H^+) ions and hydroxide (OH^-) ions as well as the ions from the dissolved compound. These hydrogen and hydroxide ions from the water can be discharged (gain or lose electrons) at the electrodes instead of the ions from the ionic compound.

Electrode	Ion from water that can be discharged	Half equation	Product
Negative	hydrogen (H^+)	$2H^+ + 2e^- \rightarrow H_2$	hydrogen gas
Positive	hydroxide (OH^-)	$4OH^- \rightarrow 2H_2O + O_2 + 4e^-$	oxygen gas

A Electrolysis of water.

Some ions are difficult to discharge. Positive ions of very reactive metals such as sodium ions (Na^+) or potassium ions (K^+) are difficult to discharge. When dissolved in water, hydrogen ions (H^+) are easier to discharge. Negative ions like sulfate ions (SO_4^{2-}) and nitrate ions (NO_3^-) are also difficult to discharge. When dissolved in water, hydroxide ions (OH^-) are easier to discharge.

Pure water is mainly made of water molecules (H_2O). Only about 0.000 01% of the molecules are broken apart into hydrogen (H^+) and hydroxide (OH^-) ions. This is why pure water is a very poor conductor of electricity. However, during electrolysis, it is often easier for more water molecules to break apart and the hydrogen ions (H^+) and/or hydroxide ions (OH^-) to be discharged than the ions from the dissolved ionic compound:

$$H_2O \rightarrow H^+ + OH^-$$

1 Why do solutions of ionic compounds conduct electricity?

2 A solution of salt, that is sodium chloride, contains sodium (Na^+) and chloride (Cl^-) ions. Name and give the formula of two other ions in the solution.

	Solution	Product
Negative electrode (positive ions)	containing ions of metals low in the reactivity series	the metal
	containing ions of metals high in the reactivity series	hydrogen
Positive electrode (negative ions)	containing halide ions (chloride, bromide, iodide)	the halogen (chlorine, bromine, iodine)
	containing other negative ions	oxygen

B Examples of the products of electrolysis of solutions using graphite (carbon) electrodes.

3 Why are ions like sodium (Na^+), potassium (K^+), sulfate (SO_4^{2-}) and nitrate (NO_3^-) ions not discharged during the electrolysis of solutions?

A good example of the electrolysis of molten and dissolved compounds is sodium chloride. In the electrolysis of molten sodium chloride, sodium and chlorine are made at the electrodes. However, in the electrolysis of sodium chloride solution, hydrogen and chlorine are formed. This is because it is easier to discharge hydrogen (H^+) ions from the water than sodium (Na^+) ions from the sodium chloride.

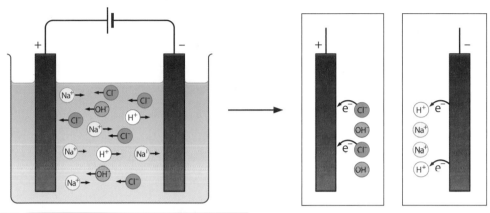

C Electrolysis of sodium chloride solution.

Solution	Formula	Ions contained	Negative electrode		Positive electrode	
			Equation	Product	Equation	Product
Lead bromide	$PbBr_2(aq)$	Pb^{2+}, Br^-, H^+, OH^-	$Pb^{2+} + 2e^- \rightarrow Pb$	lead	$2Br^- \rightarrow Br_2 + 2e^-$	bromine
Sodium chloride	$NaCl(aq)$	Na^+, Cl^-, H^+, OH^-	$2H^+ + 2e^- \rightarrow H_2$	hydrogen	$2Cl^- \rightarrow Cl_2 + 2e^-$	chlorine
Silver nitrate	$AgNO_3(aq)$	Ag^+, NO_3^-, H^+, OH^-	$Ag^+ + e^- \rightarrow Ag$	silver	$4OH^- \rightarrow 2H_2O + O_2 + 4e^-$	oxygen
Potassium sulfate	$K_2SO_4(aq)$	$K^+, SO_4^{2-}, H^+, OH^-$	$2H^+ + 2e^- \rightarrow H_2$	hydrogen	$4OH^- \rightarrow 2H_2O + O_2 + 4e^-$	oxygen

D What happens in the electrolysis of some solutions of ionic substances.

Using electrolysis

One example of a use of electrolysis of solutions is electroplating, where metals are coated with other metals by electrolysis.

E Cutlery can be plated with silver by electrolysis.

4 For each of the following solutions, state the products of electrolysis at the positive and negative electrodes:
   a copper nitrate   c sodium nitrate
   b gold chloride   d potassium bromide.

5 Calcium nitrate is an ionic substance with the formula $Ca(NO_3)_2$. It contains calcium (Ca^{2+}) and nitrate (NO_3^-) ions. In the electrolysis of a solution of calcium nitrate, for *each* electrode:
   a predict what would be formed
   b state whether the reaction is oxidation or reduction
   c write a half equation.

Using electrolysis

By the end of this topic you should be able to:

- name the products from the electrolysis of sodium chloride solution
- give uses for the products from the electrolysis of sodium chloride solution
- explain how copper is purified by electrolysis.

The electrolysis of sodium chloride solution is a major industrial process. The products have many uses. Most of the sodium chloride used for electrolysis in the UK comes from underground in Cheshire. It is extracted by pumping water underground. The sodium chloride dissolves in the water and is then pumped up to the surface. A saturated solution of sodium chloride is sometimes called brine.

A The electrolysis of salt water at Runcorn, Cheshire.

The electrolysis of the sodium chloride solution is often carried out in a membrane cell. The membrane keeps two of the products of the electrolysis, chlorine and sodium hydroxide, apart. The overall equation for the process is:

1 What are the three products in the electrolysis of sodium chloride solution?

sodium chloride + water $\longrightarrow$ sodium hydroxide + chlorine + hydrogen

$$2NaCl(aq) + 2H_2O(l) \longrightarrow 2NaOH(aq) + Cl_2(g) + H_2(g)$$

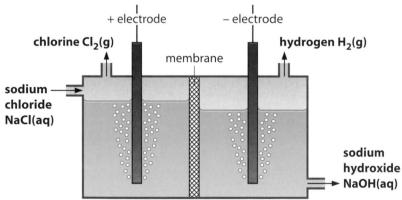

B Electrolysis of sodium chloride solution.

Product	Chlorine gas, $Cl_2(g)$	Hydrogen gas, $H_2(g)$	Sodium hydroxide solution, NaOH(aq)
Where it is formed	positive electrode	negative electrode	in solution
How it is formed	$2Cl^- \longrightarrow Cl_2 + 2e^-$	$H_2O \longrightarrow H^+ + OH^-$ $2H^+ + 2e^- \longrightarrow H_2$	left over solution contains Na^+ and OH^- ions
Uses	• to kill bacteria in drinking water • to make disinfectants • to make bleach (with NaOH) • to make PVC • to make hydrochloric acid (with H_2)	• as a fuel • to make ammonia • to make margarine • to make hydrochloric acid (with H_2)	• to make bleach (with NaOH) • to make soap and detergents • to make paper

C Products in the electrolysis of sodium chloride solution.

2 a What is formed at the negative electrode in the electrolysis of sodium chloride solution?
  b Give two uses for this product.

3 a What is formed at the positive electrode in the electrolysis of sodium chloride solution?
  b Give two uses for this product.

4 a What is formed in the solution in the electrolysis of sodium chloride solution?
  b Give two uses for this product.

5 Why is there a membrane in the electrolysis cell in the electrolysis of sodium chloride solution?

If the electrolysis of solutions is done with metal electrodes, instead of graphite, different reactions can take place at the electrodes. The purification of copper is an example of this. This process is another example of electrolysis on an industrial scale.

Copper is used in electrical cables. It has to be very pure for this use. It is purified (refined) by electrolysis of a solution containing copper (Cu^{2+}) ions using copper electrodes. A piece of impure copper is used as the positive electrode and pure copper as the negative electrode. Copper is lost from the positive electrode and pure copper formed on the negative electrode. Impurities from the impure copper collect under the positive electrode in a slime.

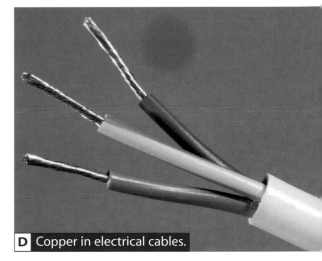

D Copper in electrical cables.

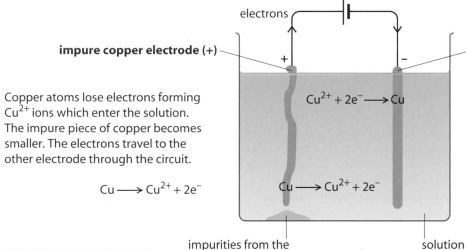

electrons

impure copper electrode (+)

pure copper electrode (–)

Copper atoms lose electrons forming Cu^{2+} ions which enter the solution. The impure piece of copper becomes smaller. The electrons travel to the other electrode through the circuit.

$$Cu \longrightarrow Cu^{2+} + 2e^-$$

$$Cu^{2+} + 2e^- \longrightarrow Cu$$

$$Cu \longrightarrow Cu^{2+} + 2e^-$$

Cu^{2+} ions in the solution are attracted to the negative electrode where they gain electrons forming copper. The pure piece of copper becomes bigger.

$$Cu^{2+} + 2e^- \longrightarrow Cu$$

impurities from the copper collect here

solution containing Cu^{2+} (aq) ions (for example, copper sulfate)

E Purification of copper.

6 Why is some copper purified by electrolysis?

7 Write a short paragraph to explain how copper is purified.

8 Produce a 2 minute presentation to explain why electrolysis is important. Use the purification of copper and the electrolysis of sodium chloride solution as examples. You could use a presentation package to illustrate your idea.

Michael and Natasha carried out an experiment on the electrolysis of copper sulfate using copper electrodes. They were trying to find out how changing the size of the electric current affects the mass of copper gained at the negative electrode. During the electrolysis, copper metal is lost from the positive electrode and copper metal is formed at the negative electrode. They set the circuit up as shown in diagram A and then measured the mass of copper gained at the negative electrode.

Their results are shown in table B.

Current (A)	Mass gained (g)			
	Test 1	Test 2	Test 3	Average
0.2	0.023	0.021	0.019	
0.4	0.039	0.042	0.041	
0.6	0.064	0.061	0.060	
0.8	0.078	0.082	0.083	
1.0	0.083	0.097	0.099	

B

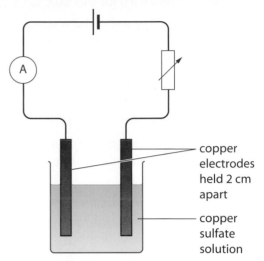

A The electrolysis of copper sulfate using copper electrodes.

copper electrodes held 2 cm apart

copper sulfate solution

1 a What was the dependent variable in this experiment? *(1 mark)*
 b What was the independent variable in this experiment? *(1 mark)*

2 a One of the results for 1.0 A is anomalous. Which result is it? *(1 mark)*
 b Do you think that this anomalous result is due to a random error or a systematic error? Explain your answer. *(2 marks)*
 c What should be done about this anomalous result? *(1 mark)*

3 Copy and complete the table by calculating the average mass gained for each current. Do not count the anomalous result in your averages. *(2 marks)*

4 a Explain why using a balance with a sensitivity of 0.1 g would not be suitable in this investigation. *(1 mark)*
 b What was the sensitivity of the balance used? *(1 mark)*

5 From the results table, which current's results show the least reliability? Is it the results at 0.2, 0.4, 0.6, 0.8 or 1.0 A? Explain your answer. *(2 marks)*

6 Which one of the following would be the best way of presenting these results: a bar chart, a line graph, a pie chart or a histogram? *(1 mark)*

7 ✎ The students said that their results showed that the mass of copper gained is directly proportional to the current. Explain what this means. Quality of written communication is important in this answer. *(3 marks)*

Total = 16 marks

Why is ammonia important?

A The ammonia made at Terra Nitrogen, Billingham is used to make fertilisers.

The world's population is about 6.3 billion and is growing. All of these people need enough food. However, there is hardly any more land that can be converted into farmland, and so we must try to grow more food on the land that we already use. This means using fertilisers. Fertilisers replace elements that have been removed from the land by growing plants. An important element that plants need is nitrogen, which is added as ammonium compounds or nitrates.

Ammonium compounds are made from ammonia, which is made by the Haber process. This process uses air and natural gas as the raw materials.

1 The list below shows things that plants need for growth.
  a Sort them into groups.
  b Explain why you have grouped them in the way that you have.

air; ammonium phosphate; ammonium sulfate; carbon dioxide; potassium chloride; sodium nitrate; sunlight; water

By the end of this unit, you should be able to:

- explain the energy transfers associated with chemical reactions
- describe what happens in reversible reactions

H
- explain what a state of equilibrium is
- describe what happens in the Haber process
- explain how the rate of a reaction can be affected by temperature, concentration or surface area of solid reactants
- describe what the collision theory and activation energy are

H
- apply these factors to reactions in equilibrium
- describe how we measure concentration
- describe how to make an insoluble salt
- describe two ways of making soluble salts
- describe how ammonia is converted into ammonium salts which are then used as fertilisers.

Energy transfers in chemical reactions

By the end of this topic you should be able to:

- explain how energy transfers are associated with chemical reactions
- describe how exothermic reactions transfer energy to the surroundings
- describe how endothermic reactions take in energy from the surroundings.

A Ammonium nitrate is a common fertiliser that increases crop growth and yield.

A **neutralisation** reaction is where an acid reacts with an alkali to form a salt. To make ammonium nitrate, the acid is nitric acid and the alkali is ammonium hydroxide:

ammonium hydroxide + nitric acid ⟶ ammonium nitrate + water

$$NH_4OH(aq) + HNO_3(aq) \rightarrow NH_4NO_3(aq) + H_2O(l)$$

As the ammonium hydroxide and nitric acid react together, heat energy is given out. We say that the reaction is **exothermic**.

When ammonium nitrate reacts with water the mixture gets colder. The reaction takes in heat energy from the surroundings. This is an **endothermic** reaction.

B In the cold pack, the reaction between ammonium nitrate and water takes heat energy away from the injured leg.

1 a What is an exothermic reaction?
  b Give an example.

2 a What is an endothermic reaction?
  b Give an example.

Chemical reactions involve transfers of energy. Many reactions give out heat energy, but sometimes they also give out sound and light energy.

Many kinds of energy are given out during a firework display.

All **combustion** reactions are exothermic. For example, in a Bunsen burner or gas oven, methane (CH_4) in the gas reacts with oxygen in the air to form carbon dioxide, water and heat energy:

methane + oxygen $\longrightarrow$ carbon dioxide + water (+ heat energy)
$$CH_4(g) + 2O_2(g) \longrightarrow CO_2(g) + 2H_2O(g) \text{ (+ heat energy)}$$

When magnesium burns in air, a lot of heat (and light) energy is produced. This is an **oxidation** reaction – the magnesium reacts with oxygen.

3 Look at photograph C. List three types of energy that are being given out.

4 a What is formed when magnesium burns in air?
  b Magnesium burning in air is an exothermic reaction. What energy transfers happen during the reaction?

Endothermic reactions are less common. One example is the **thermal decomposition** of calcium carbonate. Calcium carbonate does not decompose (split up) until it is heated to a high temperature. When the temperature is high enough, the calcium carbonate decomposes and calcium oxide and carbon dioxide are formed:

calcium carbonate $\longrightarrow$ calcium oxide + carbon dioxide
$$CaCO_3(s) \longrightarrow CaO(s) + CO_2(g)$$

This reaction only continues while it is being heated, so it is endothermic. It needs the heat energy to continue. This reaction is important in a blast furnace, and also in making cement and glass. Other metal carbonates decompose in a similar way.

Another very exothermic reaction.

5 Why is the thermal decomposition of calcium carbonate important?

6 Which two products will be formed if copper carbonate is heated?

7 What will happen if you stop heating copper carbonate?

8 Make a table to show the differences between exothermic and endothermic reactions. Give at least two examples of each.

Reversible reactions

By the end of this topic you should be able to:

- describe what happens in a reversible reaction
- explain energy transfers in a reversible reaction
- describe the test for the presence of water using white anhydrous copper sulfate.

In any chemical reaction, **reactants** are converted into **products**. When magnesium reacts with hydrochloric acid, magnesium chloride and hydrogen are formed:

$$\underbrace{magnesium + hydrochloric\ acid}_{reactants} \rightarrow \underbrace{magnesium\ chloride + hydrogen}_{products}$$

1 What are the reactants and products in these reactions:
   a methane burning in oxygen
   b magnesium burning in oxygen
   c ammonium hydroxide reacting with nitric acid? (*Hint*: Look back at topic C2.19 to help you.)

In some chemical reactions, the products can react to form the original reactants. When you heat solid ammonium chloride it decomposes to form gaseous ammonia and hydrogen chloride:

ammonium chloride $\rightarrow$ ammonia + hydrogen chloride
$$NH_4Cl(s) \rightarrow NH_3(g) + HCl(g)$$

If ammonia gas is held near hydrogen chloride gas, solid ammonium chloride is seen. The products of the first reaction have reacted to form the reactants again:

ammonia + hydrogen chloride $\rightarrow$ ammonium chloride
$$NH_3(g) + HCl(g) \rightarrow NH_4Cl(s)$$

This is called a **reversible reaction**. It can be written with arrows in both directions:

$$NH_4Cl(s) \rightleftharpoons NH_3(g) + HCl(g)$$

2 How do you show that a reversible reaction is taking place when you write an equation?

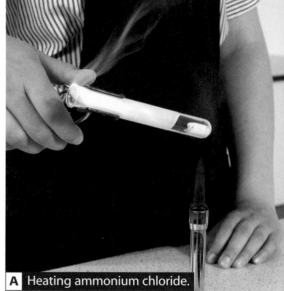

A Heating ammonium chloride.

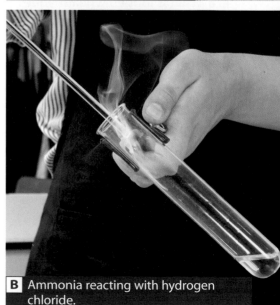

B Ammonia reacting with hydrogen chloride.

When ammonium chloride is heated it decomposes. This is an endothermic reaction. So, when ammonia and hydrogen chloride react together, heat is given out in an exothermic reaction:

$$NH_4Cl(s) \text{ (+ heat)} \rightleftharpoons NH_3(g) + HCl(g)$$

The same amount of energy is transferred in each case. This is the same for all reversible reactions. If one reaction *takes in* heat energy, the opposite will *give out* the same amount of heat energy.

This can also be seen if we heat blue hydrated copper sulfate crystals.

Blue crystals of copper sulfate contain water as well as copper sulfate. We say that they are **hydrated**. If these hydrated copper sulfate crystals are heated, they lose the water and turn into white **anhydrous** copper sulfate. When water is added to this white solid, blue hydrated crystals form again. The reaction is reversible. Heating the blue crystals is endothermic, as heat energy is taken in, so adding water to the white solid anhydrous copper sulfate produces heat energy and is exothermic:

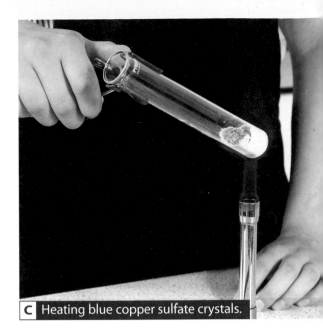

C Heating blue copper sulfate crystals.

hydrated copper sulfate (+ heat) $\rightleftharpoons$ anhydrous copper sulfate + water
(blue)   (white)

We can use the reaction between anhydrous copper sulfate and water as a test for water. If white anhydrous copper sulfate is added to a liquid and it turns blue, water is present.

3 How could you show that adding water to anhydrous copper sulfate is an exothermic reaction?

The test for water

D Testing for the presence of water.

4 Name three liquids which would turn white anhydrous copper sulfate blue.

5 List as many statements as you can about this reaction:
hydrated cobalt chloride (+ heat) $\rightleftharpoons$ anhydrous cobalt chloride + water

Equilibria

By the end of this topic you should be able to:

- explain that when a reversible reaction occurs in a closed system, equilibrium is reached
- explain that the relative amounts of the reacting substances depend on the reaction conditions.

When you heat ammonium chloride in an open test tube, all of the ammonia and hydrogen chloride formed escape into the atmosphere. They cannot react again to form ammonium chloride. However, in a sealed tube, the two gases cannot escape from the tube, so they react to form ammonium chloride again. It soon reaches a state where the ammonium chloride decomposes at the same time and the same rate as the ammonia and hydrogen chloride recombine. This is known as a state of **equilibrium** and is written ⇌.

ammonium chloride ⇌ ammonia + hydrogen chloride
$$NH_4Cl(s) \rightleftharpoons NH_3(g) + HCl(g)$$

When we talk about equilibrium reactions we refer to the **forward reaction** and the **backwards reaction**:

reactants ⇌ products

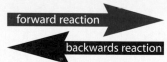

forward reaction

backwards reaction

In a **dynamic equilibrium** the forward reaction and the backwards reaction occur at the same time and rate. This means that the amounts of each do not change. The overall energy transfer is zero.

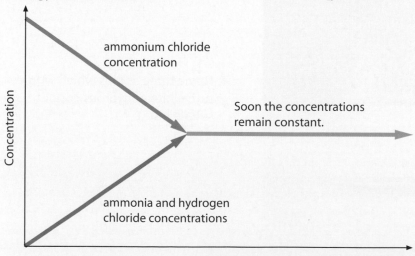

ammonium chloride concentration

Soon the concentrations remain constant.

Concentration

ammonia and hydrogen chloride concentrations

Time

B At equilibrium the amounts of reactants and products stay constant.

A Heating ammonium chloride in a sealed tube.

1 Why don't ammonia and hydrogen react again if they are produced in an open test tube?

2 What is a state of equilibrium?

We also get dynamic equilibrium between ice and water at 0°C. Ice melts to form water at the same time as the water freezes to form ice. We cannot see a change, but both of these processes occur at the same time.

An equilibrium only occurs if the reaction is in a closed system. This means that nothing can get in or out. If reactants escape, the forward reaction will occur more slowly. If products escape, the backwards reaction will occur more slowly.

C Lime kilns have been used for centuries to make calcium oxide.

When limestone (calcium carbonate) is heated in a closed tube, there is an equilibrium with the products, calcium oxide and carbon dioxide:

calcium carbonate $\rightleftharpoons$ calcium oxide + carbon dioxide

If limestone is heated in a lime kiln, the carbon dioxide formed escapes into the air. All of the calcium carbonate can decompose to form calcium oxide and carbon dioxide.

5 Why doesn't the backwards reaction take place in a lime kiln?

Changing the conditions of the reaction, changes the position of the equilibrium, and so the amount of reactants and products present. This is important in industry, where many products are made in equilibrium reactions. These industrial reactions do not give as much product as possible because they are in equilibrium. However, changing the conditions allows us to make as much of the valuable product as possible.

6 List as many statements as you can about this reaction:

sulfur dioxide + oxygen $\rightleftharpoons$ sulfur trioxide

3 Write a word equation for the equilibrium between ice and water.

4 If you found that there was 5 g of ice present after 10 minutes at 0°C, how much would there be after 20 minutes?

The Haber process

By the end of this topic you should be able to:

- give the names and sources of the raw materials for the Haber process
- explain that nitrogen and hydrogen react in a reversible reaction which does not go to completion
- describe how ammonia is removed at the end of the process, and the nitrogen and hydrogen are recycled
- appreciate the reasons for minimising energy requirements and waste.

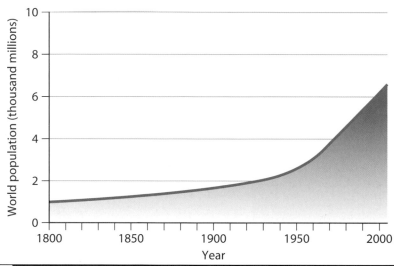

A The world's population is increasing rapidly, so we must grow more food.

All plants need nitrogen to make protein molecules. Although air contains 80% nitrogen, most plants cannot use it directly. We have to use fertilisers which contain nitrogen in the form of ammonium compounds or nitrates, to replace nitrogen compounds that have either been used up by plants or have been washed away by rain water.

1 a How is the world's population changing?
  b Explain why this change means that more fertilisers are needed.

2 Why is ammonium nitrate (NH_4NO_3) such a good fertiliser?

Ammonia is made by using the **Haber process**. In this process hydrogen and nitrogen react together in a reversible reaction:

nitrogen + hydrogen $\rightleftharpoons$ ammonia
$$N_2(g) + 3H_2(g) \rightleftharpoons 2NH_3(g)$$

The nitrogen comes from the fractional distillation of air. The hydrogen comes from natural gas and other sources such as oil.

3 a What are the raw materials in the Haber process?
  b Where does each raw material come from?

B Part of the Haber plant at Billingham.

The nitrogen and hydrogen are purified and then reacted together. Some of the hydrogen and nitrogen combine to form ammonia. However, this process is reversible, so there is always a mixture of hydrogen, nitrogen and ammonia. The amount of ammonia formed depends on the conditions of temperature and pressure used. The conditions normally used are a temperature of about 450°C and a pressure of 200 atmospheres. A catalyst of iron helps to speed up the process.

At the end of the reaction, the ammonia formed is cooled. It liquefies and can be removed. The conditions used in the Haber process give a yield of about 15% ammonia. The hydrogen and nitrogen that have not reacted are not wasted, but are recycled and reused.

4 What are the conditions for the Haber process?

5 What does the iron catalyst do?

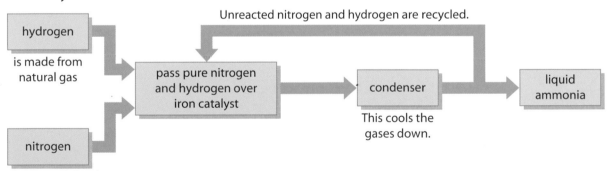

C About 150 million tonnes of ammonia are made worldwide every year.

Sustainable industrial processes

Making ammonia uses up about 1% of the world's energy production. It is therefore important for sustainable development as well as economic reasons to reduce the energy used and wasted in this industrial process. We can do this by altering the conditions of the equilibrium.

The majority of ammonia made by this process is used to make fertilisers. Some is also converted into nitric acid:

ammonia + oxygen → nitric acid + water
$$NH_3(g) + 2O_2(g) \rightarrow HNO_3(aq) + H_2O(l)$$

6 Why are the unused nitrogen and hydrogen recycled?

7 Why do we want to save energy?

8 Write an encyclopaedia entry titled 'The Haber process'.

Rates of reaction

By the end of this topic you should be able to:

- show how the rate of a reaction can be found by measuring the amount of reactant used or the amount of product formed over time
- interpret graphs showing the amount of product formed (or reactant used up) with time.

Manufacturers want to make as much of their product as possible in the shortest amount of time. They want reactions that are fast, but safe, and also have a high yield.

1 Why do manufacturers want fast reactions?

2 Give an example of a very fast reaction

3 Give an example of a very slow reaction.

The speed of a reaction is called the **rate of reaction**. It can be shown as:

A This reaction would be far too fast in industry!

$$\text{rate of reaction} = \frac{\text{amount of reactant used or amount of product formed}}{\text{time}}$$

Graph B shows how the reactants are used up and the products are formed during a reaction.

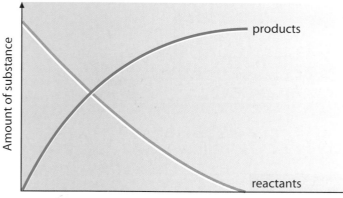

B How the amounts of reactant and product change during a chemical reaction.

There are several ways of measuring the rate of a reaction. If the reaction produces a gas, we can measure the rate of reaction by measuring the volume of gas given off or loss in mass of the chemicals during the reaction.

Calcium carbonate reacts with hydrochloric acid to form carbon dioxide gas:

calcium carbonate + hydrochloric acid ➞ calcium chloride + carbon dioxide + water

$$CaCO_3(s) \quad + \quad 2HCl(aq) \quad \longrightarrow \quad CaCl_2(aq) \quad + \quad CO_2(g) \quad + H_2O(l)$$

Diagram C shows how you can measure the volume of gas given off.

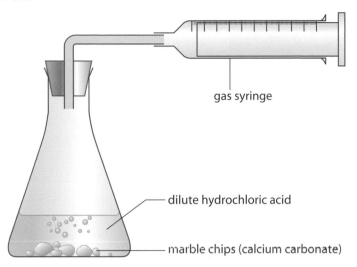

gas syringe

dilute hydrochloric acid

marble chips (calcium carbonate)

C Apparatus used to measure the volume of gas given off.

Graph D shows how much gas was given off in the same reaction. The graph is steepest at the start. The rate of the reaction is the greatest here as there is the most acid and marble chips available to react. As the reactants are used up there are less of them. The reaction gets slower and the graph flattens off. When the graph is totally flat, the reaction has finished because one or both of the reactants has been used up.

5 Marble chips are usually left in the flask at the end of this reaction. Which reactant has been used up?

The reaction can also be followed by finding how the mass changes. Diagram E shows the same experiment carried out on a balance. The loss of mass can be measured every minute until the reaction finishes.

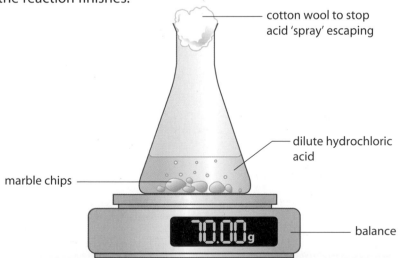

cotton wool to stop acid 'spray' escaping

dilute hydrochloric acid

marble chips

balance

E Apparatus used to measure the change in mass of chemicals in the flask.

4 Gas can also be collected in an upside-down measuring cylinder in a bowl of water. Suggest why it is better to use a gas syringe.

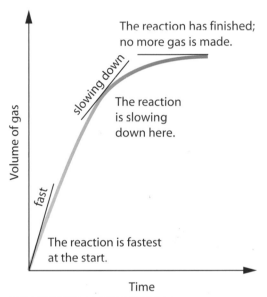

The reaction has finished; no more gas is made.

slowing down

The reaction is slowing down here.

fast

The reaction is fastest at the start.

Time

D Volume of gas given off.

6 How will you know that the reaction is finished?

7 Sketch a graph, with labelled axes, showing how the mass of the apparatus would change during this reaction. Remember that the carbon dioxide produced is leaving the reaction flask.

8 Magnesium reacts with hydrochloric acid to form magnesium chloride and hydrogen gas. Draw a poster that shows everything that you could do to measure the rate of this reaction.

Temperature and reaction rates

By the end of this topic you should be able to:

- explain that particles need to collide before they can react
- explain what activation energy is
- describe how increasing the temperature increases the speed of the particles so that they collide more frequently and with more energy, and this increases the rate of the reaction.

All chemicals consist of tiny particles. Different chemicals can only react when these particles bump into each other. If they hit each other hard enough, they will react. This is called the **collision theory**. The minimum amount of energy that these particles must have to react is called the **activation energy**. Only those particles with energy greater than the activation energy will react, the others won't.

1 What is the collision theory?

2 What is activation energy?

A Food lasts longer if it is kept cold.

Fridges slow down the chemical reactions that occur when food 'goes off'. Just as lowering the temperature slows down reactions, heating speeds them up. This can be explained using the collision theory. As the temperature increases, particles move around faster and are much more likely to collide with each other. They also gain energy and hence more of them have energy greater than the activation energy. A combination of more collisions and increased energy leads to a faster reaction.

3 Why does cooling a reaction slow it down?

The effect of increasing temperature on reaction rates

You can show the effect of increasing the temperature experimentally. The experiment with marble chips and acid can be repeated at different temperatures. Graph B shows some results. You can see that when the temperature of a reaction is increased, more gas is formed in the same time. The reaction is faster.

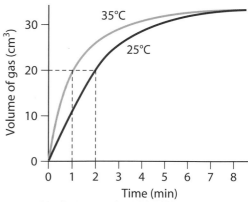

B Effect of increasing the temperature on reaction rates.

4 Name two variables that should be kept constant during this reaction to make the comparison fair.

The disappearing cross

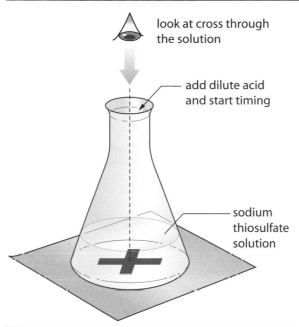

look at cross through the solution

add dilute acid and start timing

sodium thiosulfate solution

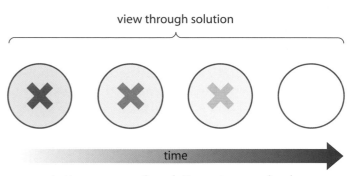

view through solution

time

As time goes on, the solution gets more cloudy. The cross 'disappears'.

C Apparatus for the 'disappearing-cross' reaction.

Diagram C shows another experiment to find the effect of temperature, using the 'disappearing-cross' method. A solution of sodium thiosulfate reacts with hydrochloric acid. Solid sulfur is produced and the reaction mixture goes cloudy:

sodium thiosulfate + hydrochloric acid → sodium chloride + sulfur + sulfur dioxide + water

$$Na_2S_2O_3(aq) + 2HCl(aq) \rightarrow 2NaCl(aq) + S(s) + SO_2(g) + H_2O(l)$$

We can follow this reaction by finding how long it takes for the cross to disappear. The faster the reaction, the faster the cross will disappear. Table D shows sample results for the time for the cross to disappear at different temperatures.

Temperature of reaction mixture (°C)	Time for cross to disappear (seconds)
20	360
30	180
40	91
50	46
60	23

D Time for the cross to disappear at different temperatures.

5 Why does the reaction of sodium thiosulfate and hydrochloric acid go cloudy?

6 Plot a graph showing how the time taken for the cross to disappear changes as you increase the temperature. Plot time on the *y*-axis and temperature on the *x*-axis.

7 Predict how long the reaction would take at 45 °C.

8 Explain why chemical reactions take place when food is cooked in an oven.

P

D

P

P

Surface area and reaction rates

By the end of this topic you should be able to:

- explain how the rate of a chemical reaction increases if the solid reactants have a greater surface area
- relate rate of reaction and surface area to collision theory.

If you cut a potato into small pieces it will cook much faster than a whole potato. The smaller the pieces of potato, the faster the reaction is. This is because more surface is exposed to the boiling water. It is the same with chemical reactions.

1 Which variables should be kept constant in the reaction shown in graph A to make it a fair test?

2 Which reaction used marble with the largest surface area?

3 Which reaction finished first?

4 How would you explain these results?

The rate of the reaction does not change the total amounts of the products. It only changes the speed at which they are formed.

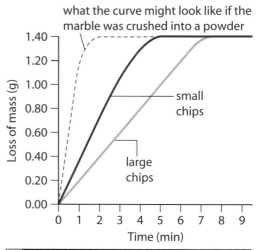

Rate-of-reaction graphs

what the curve might look like if the marble was crushed into a powder

small chips

large chips

A This graph shows what happens if you react hydrochloric acid with marble chips and powdered marble.

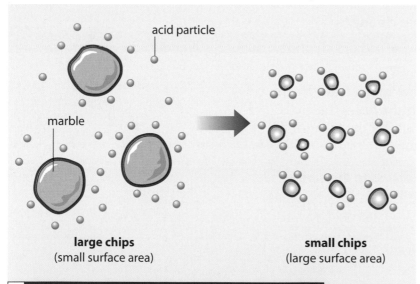

acid particle

marble

large chips
(small surface area)

small chips
(large surface area)

B The smaller the marble chips, the faster the reaction.

Collision theory can be used to explain the effect of surface area. The smaller the marble chips, the larger the surface area for the same mass of marble. The larger the surface area, the greater the number of collisions that can happen between the acid and the marble chips at the same time. This means that the reaction occurs faster.

5 How does collision theory explain the effect of changing surface area on a reaction rate?

Rate of reaction – magnesium and hydrochloric acid

C Magnesium reacting with hydrochloric acid.

Look at the photographs in C. You can see that magnesium powder reacts with hydrochloric acid much more rapidly than magnesium ribbon does. We can show this experimentally. Magnesium ribbon is placed in a conical flask attached to a gas syringe. Hydrochloric acid is added and the volume of hydrogen formed is measured every 10 seconds. The experiment is then repeated using powdered magnesium. Table D shows the results.

Time (s)	Volume of hydrogen produced from magnesium ribbon (cm³)	Volume of hydrogen produced from magnesium powder (cm³)
0	0	0
10	22	52
20	40	68
30	56	77
40	68	82
50	76	84
60	82	86
70	85	86
80	86	86
90	86	86
100	86	86

D Results with powdered and ribbon magnesium.

6 Plot a graph showing these results, using axes like these.

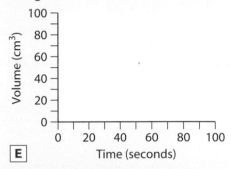

E

7 Why is the final volume of hydrogen the same in both experiments?

8 Why do the graphs flatten off near the end of the reaction?

9 a Which reaction produced the most hydrogen in the first 30 seconds?
   b What is the average rate of production of hydrogen in cm³ per second for each reaction during this time?

10 Explain the difference in the results.

11 Hydrogen is the lightest gas that there is. Why can't this experiment be done by finding mass loss?

12 Explain why powdered charcoal burns much more quickly than a lump of coal.

Concentration and reaction rates

By the end of this topic you should be able to:

- explain how increasing the concentration of reactants in solution affects the rate of a reaction
- explain how increasing the pressure of gases increases the rate of a reaction.

Concentrated and dilute solutions

The **concentration** of a solution tells us how much **solute** is dissolved in the water. If there is a large amount of solute in a small amount of water, the solution is **concentrated**. If there is a lot of water and little solute, the solution is **dilute**.

1 spatula of
copper sulfate
added to water

3 spatulas of
copper sulfate
added to water

A A dilute solution and a concentrated solution.

Chemical reactions can be carried out using different concentrations of solutions. If you use a more concentrated solution of one of the reactants, the reaction is faster. This is because there are more particles present, and so they are more likely to collide with the other particles in the mixture. The more successful collisions there are, the faster the reaction is.

Magnesium and different concentrations of acid

Solution	Volume of acid (cm³)	Volume of water (cm³)	Time to collect 20 cm³ of gas (seconds)
A	10	40	200
B	20	30	100
C	30	20	67
D	40	10	50
E	50	0	40

B Results of an experiment reacting the same mass of magnesium with different concentrations of acid.

1 A student made two different solutions of sodium chloride:
solution A: 2 g dissolved in 200 cm³ water
solution B: 4 g dissolved in 500 cm³ water
   a Calculate the concentrations of these solutions in grams per cm³.
   b Which solution has the greater concentration?

Look at table B.

2 a Which solution is the most concentrated?
 b Which solution gives the fastest reaction?

3 Why didn't the people doing the experiment use a solution containing no acid?

4 Plot a graph of volume of acid used in cm³ against time to collect 20 cm³ of gas. Use axes like these.

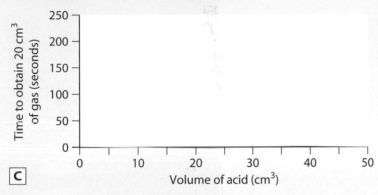

C

5 Use your graph to find the time needed to collect 20 cm³ of gas when the volume of acid used was:
 a 15 cm³
 b 45 cm³.

6 Which variables should be kept constant during this reaction?

If we change the pressure of a reaction where the reactants are either solids or liquids, there is no effect on the rate of the reaction. However, if the reactants are gases, the rate increases as the pressure increases. This is because if you have a fixed amount of gas and increase the pressure, the gas is squeezed into a smaller volume. This means that the particles are forced closer together and are much more likely to collide with one another.

same number of particles squeezed into smaller volume

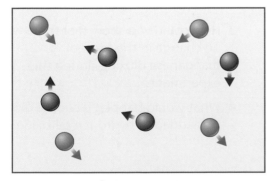

lower pressure

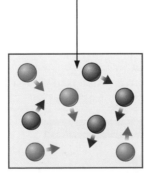

higher pressure

D The chances of collision between gas particles are much greater at a higher pressure.

7 Sulfur dioxide gas reacts with oxygen gas to form gaseous sulfur trioxide:

$$2SO_2(g) + O_2(g) \longrightarrow 2SO_3(g)$$

 a How would increasing the pressure change the rate of the reaction?
 b Explain your answer in terms of the collision theory.

8 Explain how increasing the concentration of reactants increases the rate of a reaction. Sketch a graph to illustrate your answer.

Catalysts

By the end of this topic you should be able to:

- explain what a catalyst is
- evaluate the advantages and disadvantages of using a catalyst in industrial processes.

Many chemical reactions used in industry are very slow. If the products are made slowly the manufacturers may lose money. Some chemical reactions can be speeded up using a **catalyst**. This is a chemical that is present in the reaction, but is not used up and remains unchanged at the end. It can be removed and reused. Each catalyst is **specific**, and only works on a particular reaction. Processes with several steps may need several different catalysts.

1 What is a catalyst?

2 Why do you think that catalysts are often used as a wire mesh?

We can show the effect of a catalyst by looking at this reaction:

hydrogen peroxide $\longrightarrow$ water + oxygen

$$2H_2O_2(aq) \longrightarrow 2H_2O(l) + O_2(g)$$

Hydrogen peroxide solution only decomposes very slowly by itself. If a catalyst called manganese dioxide is added, bubbles of oxygen can be seen. The rate of reaction has increased. The manganese dioxide is specific for this reaction as other metal oxides have no effect.

A Catalytic converters in cars turn poisonous exhaust gases into harmless ones. The catalysts speed up the reactions.

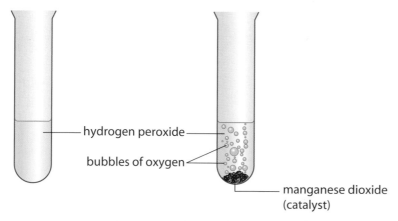

hydrogen peroxide

bubbles of oxygen

manganese dioxide (catalyst)

B Effect of adding manganese dioxide to hydrogen peroxide.

3 How could you show that there was no change in the mass of manganese dioxide during this experiment?

4 What would happen if copper oxide was added to hydrogen peroxide?

Comparing the effect of catalysts

We can also compare the effect of catalysts by measuring the amount of oxygen produced after different times. Diagram C shows the apparatus used. Table D shows some sample results.

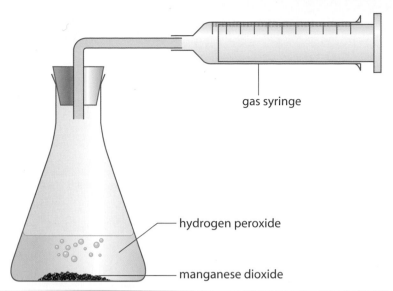

gas syringe

hydrogen peroxide

manganese dioxide

C Apparatus to measure the gas produced when hydrogen peroxide decomposes.

	Time (seconds)									
	0	**10**	**20**	**30**	**40**	**50**	**60**	**70**	**80**	**90**
Volume of oxygen (cm³)	0	11	20	27	32	36	38	39	40	40

D Volume of oxygen produced as hydrogen peroxide decomposes.

Catalysts work because they lower the activation energy of reactions. This means that more reactant particles will have enough energy to react at a lower temperature.

Catalysts are important in industry because they help to speed up reactions without having to heat the reactants to high temperatures. This saves a lot of money. Table F gives some examples of catalysts used in industry.

Industrial process	Catalyst used
Haber process to make ammonia	iron
Making sulfuric acid	vanadium oxide
Making nitric acid	platinum and rhodium
Making margarine	nickel

F Some of the catalysts used in industry.

The disadvantage of some catalysts include:
- they may become poisoned by impure reactants. They then need purifying before being re-used
- some are very expensive.

6 List three industrial processes that use catalysts.

7 For catalysts used in industry, list:
  a two advantages
  b two disadvantages.

5 a Draw a graph of the results in table D.
  b Sketch on the graph the results that would be obtained if no catalyst was used.

E Nickel is a catalyst used to obtain the hydrogen needed for the Haber process.

8 Write an encyclopaedia entry titled 'Catalysts'. Give some examples in your answer.

Making as much ammonia as possible

By the end of this topic you should be able to:

- explain how changing the temperature changes the position of an equilibrium
- explain how changing the pressure affects the position of an equilibrium
- explain how these factors, together with rate, determine the conditions used in industry
- evaluate how these reaction conditions are chosen to produce a reasonable yield of ammonia quickly.

A Industrial processes are carried out on an enormous scale.

Many industrial processes do not give as much product as you might expect because they are in equilibrium. However, if we change the conditions of temperature and pressure, we can increase the **yield** (amount) of product.

If one of the reactions in a reversible reaction is endothermic, the reverse reaction is exothermic. It is the same with reversible reactions that are in equilibrium. If we heat an equilibrium reaction, the endothermic reaction which takes in energy will work faster. Therefore if the temperature is raised, the yield from the endothermic reaction increases and the yield from the exothermic reaction decreases. Similarly, if the temperature is lowered, the yield from the endothermic reaction decreases and the yield from the exothermic reaction increases.

B The Haber process needs a lot of electricity to heat the reaction.

Changing the pressure of a reaction involving gases can also affect yield. An increase in pressure will favour the reaction that produces the least number of molecules as shown by the symbol equation for the reaction.

1 How does increasing the temperature affect the yield from:
   a an endothermic reaction
   b an exothermic reaction?

2 Will changing the pressure affect an equilibrium which only contains solutions? Explain your answer.

Both temperature and pressure must be considered when finding the best conditions for the Haber process. We also have to consider their effect on the rates of the reactions involved.

nitrogen + hydrogen $\rightleftharpoons$ ammonia

$$N_2(g) + 3H_2(g) \rightleftharpoons 2NH_3(g)$$

- The reaction which makes ammonia is exothermic. This means that a low temperature would produce more ammonia. However, if the temperature of a reaction is low, the rate of the reaction is slow. A compromise temperature of 450 °C is used. Anything higher would produce too little ammonia, anything lower would work too slowly.
- The reaction to make ammonia causes a decrease in the number of molecules of gas. Therefore a high pressure produces more ammonia. However, very high pressures are dangerous and expensive to produce. A pressure of 200 atmospheres is used. Increasing the pressure also speeds up the reactions.
- A catalyst of iron is used to speed up both the forward and backwards reactions.

These conditions therefore produce a reasonable yield of ammonia quickly.

3 Why is the temperature of 450 °C used in the Haber process?

4 Why is the pressure not raised above 200 atmospheres?

5 Why is an iron catalyst used?

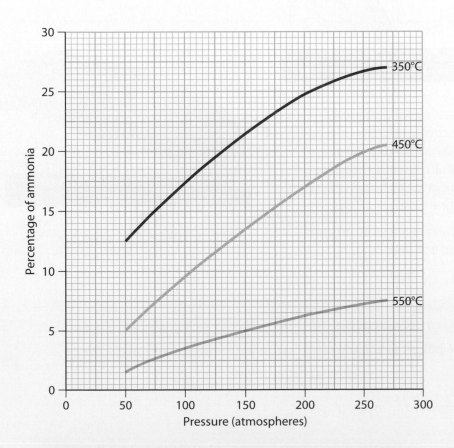

C The effect of temperature and pressure on the yield of ammonia.

Look at graph C.

6 At 450 °C, what is the yield of ammonia at:
   a 100 atmospheres
   b 300 atmospheres?

7 At 200 atmospheres, what is the yield of ammonia at:
   a 350 °C
   b 450 °C?

8 Explain why the conditions of the Haber process have been chosen.

Gases and solutions

H **By the end of this topic you should be able to:**

- explain that equal volumes of gases at the same temperature and pressure have the same number of molecules
- express concentration as moles per cubic decimetre (mol/dm³)
- explain that equal volumes of solutions of the same concentration have equal numbers of moles of solute.

A 'Avogadro said that there are 25 thousand million million million molecules in every litre of gas. I don't know how he counted them!'

Gases contain molecules which are very far apart. In 1811, an Italian called Amedeo Avogadro said that equal volumes of gases at the same temperature and pressure contain equal numbers of particles. This means that if you have 100 cm³ of oxygen at a given temperature and pressure, it contains exactly the same number of particles as there are in 100 cm³ of hydrogen, carbon dioxide or any other gas at the same conditions.

One mole of gas occupies 24 dm³ at room temperature and pressure. This is often called the **molar volume**.

B Helium atoms are very light and far apart.

$$1 \text{ dm}^3 = 1000 \text{ ml} = 1 \text{ litre}$$

1 What did Avogadro state?

2 Which contains more particles at the same temperature and pressure, 200 cm³ of hydrogen or 400 cm³ of nitrogen?

3 At room temperature and pressure, what is the volume of:
   a 0.5 moles of sulfur dioxide gas
   b 2 moles of helium gas?

4 At room temperature and pressure, how many moles of gas are there in:
   a 6 dm³ of carbon dioxide
   b 240 dm³ of methane?

The concentration of a solution is measured as the number of moles of solute dissolved in a total volume of 1 dm³ of solution. The units are moles per dm³. This is often shortened to mol/dm³.

Concentration can be calculated if we know the moles of the solute and the volume of the solution:

$$\text{concentration (mol/dm}^3) = \frac{\text{moles of solute}}{\text{volume (dm}^3)}$$

We can also relate these by using a formula triangle:

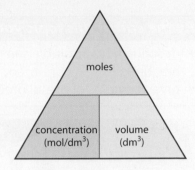

> **Example**
>
> 0.20 dm³ of a solution of sodium hydroxide contains 0.20 moles of sodium hydroxide. What is its concentration?
>
> $$\text{concentration} = \frac{0.20}{0.20} \text{ mol/dm}^3$$
> $$= 1.0 \text{ mol/dm}^3$$

If we know the concentration and volume of the solution, we can calculate the moles of solute using this equation:

moles of solute = concentration (moles/ dm³) × volume (dm³)

> **Example**
>
> A solution of calcium chloride has a concentration of 0.4 mol/dm³. How much calcium chloride is there in 0.50 dm³ of solution?
>
> 0.4 × 0.50 = 0.2 moles

This also means that equal volumes of solution of the same concentration have equal numbers of moles of solute.

5 Copy and complete table C.

Compound	Moles	Volume (dm³)	Concentration (moles/dm³)
Potassium iodide	2.0	4.00	
Sugar	0.5	0.25	
Calcium chloride		0.10	0.2
Copper sulfate		0.50	1.0

6 Tom made two different solutions of copper nitrate (see photograph D):
   solution A: 0.2 moles dissolved in 0.20 dm³ water
   solution B: 0.4 moles dissolved in 0.50 dm³ water
   a Calculate the concentrations of these solutions.
   b Which solution had the greater concentration?

7 Write notes to remind you how to work out concentrations of solutions.

Insoluble salts

By the end of this topic you should be able to:

- describe how insoluble salts are made by precipitation reactions
- give examples of useful precipitation reactions.

A The paint for these lines contains insoluble yellow lead chromate.

negative

B Sunlight breaks up insoluble silver chloride to form silver and chlorine. The black silver forms the dark parts on the negative.

Soluble salts dissolve in water. However, some salts like silver chloride and lead chromate do not dissolve in water. They are **insoluble**.

Insoluble salts can be made by mixing together **solutions** of two soluble salts. For example, silver chloride can be made by mixing solutions of sodium chloride and silver nitrate. White insoluble silver chloride immediately appears in the solution. This is called a precipitate, and this type of reaction is called a **precipitation reaction**. The insoluble silver chloride can then be filtered, washed with water and dried in a hot oven.

sodium chloride + silver nitrate ➤ silver chloride + sodium nitrate

$NaCl(aq) + AgNO_3(aq) \rightarrow AgCl(s) + NaNO_3(aq)$

1 a Name two useful insoluble salts.
 b Explain how they are useful.

2 Why must the salt in road marker be insoluble?

silver nitrate solution

sodium chloride solution

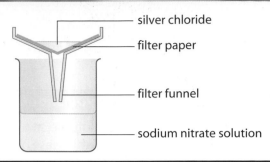

silver chloride

filter paper

filter funnel

sodium nitrate solution

C A solid precipitate of silver chloride forms when solutions of sodium chloride and silver nitrate are mixed.

D The silver chloride formed is separated by filtering it. The filtrate is sodium nitrate solution.

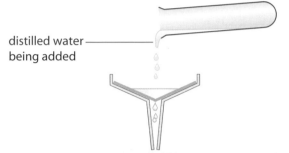

distilled water being added

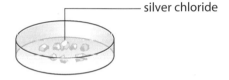

silver chloride

E The silver chloride is washed with water to remove any left-over solutions.

F The residue of silver chloride is left to dry in a dark place.

3 Look at diagrams C to F.
  a Suggest why the silver chloride made should not be dried in bright sunlight.
  b Which ions are in the filtrate in diagram D?

4 a Describe how you would make dry lead chromate ($PbCrO_4$) from lead nitrate ($Pb(NO_3)_2$) and potassium chromate (K_2CrO_4).
  b Write a balanced symbol equation for this reaction.

H

Using precipitation reactions

Precipitation reactions can also be used during water and effluent treatment. Water from a reservoir is first filtered to remove large bits of solid like fish and twigs. Aluminium sulfate and calcium hydroxide are then added to the water. These combine together in a precipitation reaction to form insoluble aluminium hydroxide:

aluminium sulfate + calcium hydroxide → aluminium hydroxide + calcium sulfate

This precipitate is then used to absorb many of the molecules in water that can cause colour or unpleasant tastes.

A precipitation reaction is also used in making water softer. Hard water (which does not form a lather with soap) contains calcium ions. These are removed by reacting the water with sodium carbonate. Insoluble calcium carbonate is formed and the hardness is removed from the water.

5 A lot of hard water contains dissolved calcium sulfate. Write a word equation showing its reaction with sodium carbonate.

6 Write a paragraph explaining:
  a what happens during a precipitation reaction
  b why precipitation reactions are so useful.

Making soluble salts

By the end of this topic you should be able to:

- describe how to make a soluble salt from an acid and either a metal or an insoluble base
- explain that the salt formed depends on the acid and base used
- describe how salt solutions can be crystallised to produce the solid salt.

Soluble salts dissolve in water.

Using soluble salts

A The salt used on food is sodium chloride. You can taste it because it dissolves in the water on your tongue.

B Copper sulfate dissolves in water and its solution is used as a spray to protect grapevines from insects.

1 a Give two examples of soluble salts.
  b Explain what each salt is used for.

Soluble salts can be made in different ways. The most common way is by reacting an acid with a metal or a metal oxide:

metal + acid ⟶ salt + hydrogen
metal oxide + acid ⟶ salt + water

You can make the salt that you want by choosing the metal or metal oxide and the acid that you use.

Many metals are not suitable. Some, like potassium, are too reactive, while some, like copper, are not reactive enough. Metals that are suitable include magnesium, zinc and iron.

More salts can be made by reacting the acid with metal oxides. For example, you can make copper sulfate by adding copper oxide to sulfuric acid. You can't use copper like this to make copper sulfate as copper is too unreactive.

Different acids give different salts:
- sulfuric acid gives sulfate salts
- hydrochloric acid gives chloride salts
- nitric acid gives nitrate salts.

Many metal oxides do not dissolve in water. If you add the insoluble solid to the acid until solid is left over, you will know that all the acid has reacted.

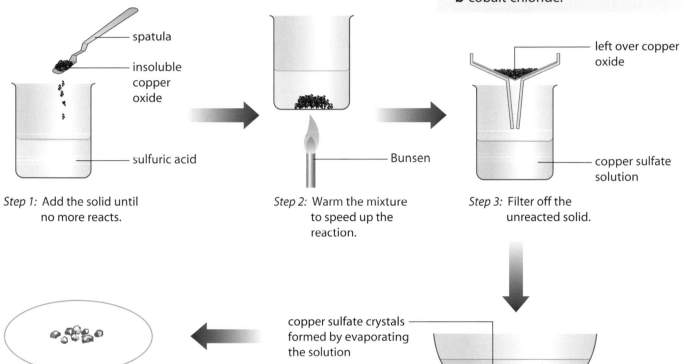

Step 1: Add the solid until no more reacts.

Step 2: Warm the mixture to speed up the reaction.

Step 3: Filter off the unreacted solid.

Step 5: Filter off the crystals and leave to dry on filter paper.

Step 4: Leave to cool. This is **crystallisation**. The slower the cooling, the larger the crystals.

C How to produce copper sulfate crystals.

The equation for the reaction to produce copper sulfate is:

copper oxide + sulfuric acid $\longrightarrow$ copper sulfate + water

$$CuO(s) + H_2SO_4(aq) \longrightarrow CuSO_4(aq) + H_2O(l)$$

4 Look at diagram C.
   a Why was the mixture heated in Step 2?
   b What would you see in Step 3 if green insoluble copper carbonate was used instead of copper oxide?
   c Why is the solution above the crystals in Step 4 still blue?

5 Write two word equations to show two methods of making zinc nitrate ($Zn(NO_3)_2$)

H 6 Write balanced symbol equations for the reactions in question **5**.

2 a Silver is a very unreactive metal. Would it react with acid?
   b If you wanted to make silver nitrate, what would you use?

3 Name the metal oxide and the acid you would use to make:
   a copper nitrate
   b cobalt chloride.

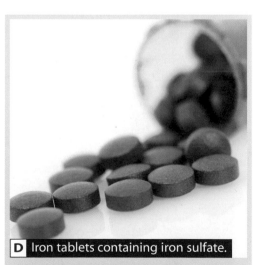

D Iron tablets containing iron sulfate.

7 Describe a method to make the iron sulfate shown in photograph D.

Bases and alkalis

By the end of this topic you should be able to:

- explain what bases and alkalis are
- describe the effect of H⁺ ions on OH⁻ ions in solution
- use the pH scale as a measure of how acidic or alkaline a solution is.

Common **acids** include hydrochloric acid (HCl), sulfuric acid (H_2SO_4) and nitric acid (HNO_3). Acids are also found in food and drinks such as orange juice.

When acids dissolve in water, they form **hydrogen ions** (H⁺). The hydrogen ions make the solution acidic. All acids have these properties:

- their pH is below 7
- they react with metals, such as magnesium, to form a salt and hydrogen
- they react with bases to form a salt and water
- they react with carbonates to form a salt, carbon dioxide and water
- they turn litmus paper red.

Bases are the opposite of acids. All metal oxides and hydroxides are bases. Some bases dissolve in water to form **alkalis**. Common alkalis are sodium hydroxide (NaOH), calcium hydroxide ($Ca(OH)_2$) and ammonia solution(NH_4OH).

All alkalis contain **hydroxide ions** (OH⁻). The hydroxide ions make the solution alkaline. Alkalis have the following properties:

- their pH is greater than 7
- they neutralise acids to form a salt and water
- they turn litmus paper blue.

A All of these contain alkalis.

1 What would be formed if you added:
   a nitric acid to magnesium
   b hydrochloric acid to sodium carbonate?

2 What colour would orange juice turn litmus paper?

3 What colour would toothpaste turn litmus paper?

4 How would you use litmus paper to distinguish between acids and alkalis?

5 Which ions do the following contain:
   a acids
   b alkalis?

We can safely swallow some acids like the citric acid in oranges, and some alkalis like antacids. These acids and alkalis are weaker than others. Hydrochloric acid and sodium hydroxide are far too dangerous to swallow. An acid like hydrochloric acid forms lots of hydrogen ions when added to water and is called a **strong acid**. Sodium hydroxide forms lots of hydroxide ions in water and is a **strong alkali**. Acids and alkalis that do not form many ions are called **weak acids** and **weak alkalis**.

6 a What is the difference between a weak and a strong acid?
   b What is the difference between a strong and a weak alkali?

Litmus only tells us if a solution is an acid or an alkali. It does not tell us how strong the acid or alkali is. We use the **pH scale** to do this.

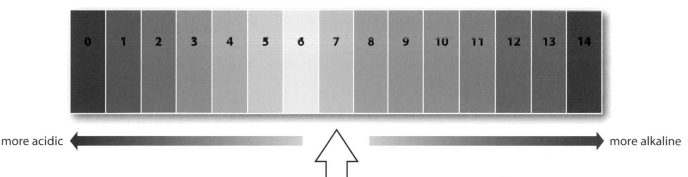

more acidic ← → more alkaline

neutral

B The pH scale.

Universal indicator changes to different colours depending on how strong the acid or alkali is. These different colours are put on a scale and given a number called the pH number. The stronger the acid, the lower the pH number. The stronger the alkali, the higher the pH number. Water is neutral and has a pH of 7.

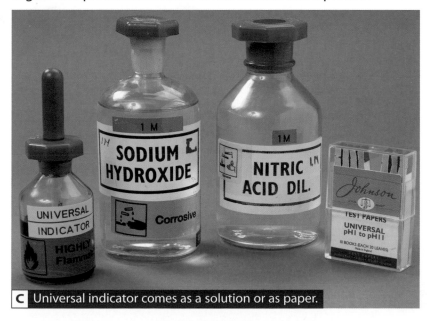

C Universal indicator comes as a solution or as paper.

7 What colour will universal indicator go with each of the two solutions shown in photograph C?

8 Write a definition for all the words in bold in this topic.

Neutralisation reactions

By the end of this topic you should be able to:

- name the ions that react together in a neutralisation reaction
- describe how to use an indicator to follow neutralisation
- represent neutralisation as an ionic equation.

We can also make soluble salts in a neutralisation reaction, by reacting an acid with an alkali:

acid + alkali ⟶ salt + water

For example:

hydrochloric acid + sodium hydroxide ⟶ sodium chloride + water

$$HCl + NaOH \rightarrow NaCl + H_2O$$

During a neutralisation reaction, the hydrogen ions from the acid react with the hydroxide ions from the alkali. This can be shown by the **ionic equation**:

$$H^+(aq) + OH^-(aq) \rightarrow H_2O(l)$$

This is the same equation whatever the acid or alkali used. The acid provides the hydrogen ions, the alkali the hydroxide ions.

To make a salt so that no acid or alkali is left over, we need to know the exact volumes of the solutions that react together. We find out these quantities using a **titration**. We use an **indicator** which changes colour at the **end point** when the acid and alkali have completely reacted.

1 What is a neutralisation reaction?

2 Why do we use an indicator in a titration?

bromothymol blue

methyl orange

phenolphthalein

acid

alkali

acid

alkali

acid

alkali

A Indicators like these are needed to find the end point of a titration.

burette

pipette

B Doing a titration.

Titrations

Titrations use specialised pieces of apparatus, a **pipette** and a **burette**. A pipette accurately measures out either 10 cm³ or 25 cm³ of a liquid. A burette can hold 50 cm³ of liquid, which can be run out, 0.1 cm³ at a time, so an accurate volume can be measured.

Step 1: Use a pipette to put a measured volume of alkali into a conical flask. Put the flask on a white tile.

Step 2: Add an indicator to the alkali in the flask.

Step 3: Fill the burette with acid.

Step 4: Write down the reading on the burette, and then open the tap to let out 1 cm³ of acid into the flask. Swirl the flask gently.

Step 5: Add another 1 cm³ of acid from the burette, and swirl again. Continue adding 1 cm³ at a time until the indicator changes colour from yellow to red.

C Indicator has been added to the alkali.

D The indicator goes red at the end-point.

Step 6: Write down the reading on the burette, and work out the volume of acid added.

Step 7: You can now make the salt by using the same quantities of acid and alkali, but leaving out the indicator. Pour the neutralised mixture into an evaporating dish and evaporate the solution to dryness. Crystals of the salt will form.

3 When looking at colour changes, why is it useful to use a white tile?

4 Why is the flask swirled in Step 4?

5 Why was the indicator left out when the salt was made in Step 7?

6 What alkali and acid would you use to make:
   a sodium sulfate   **b** potassium chloride?

7 Write word equations for the two reactions in question **6**.

H **8** Write balanced symbol equations for the two reactions in question **6**.

9 Describe how you would make potassium sulfate from potassium hydroxide and sulfuric acid.

Making fertilisers

By the end of this topic you should be able to:

- describe what is produced when ammonia dissolves in water
- explain how ammonia solution is used to make ammonium salts
- give an example of the use of ammonium salts.

Alkalis are formed when metal hydroxides dissolve in water because they form hydroxide (OH⁻) ions. When ammonia dissolves in water to form ammonia solution, some of the ammonia reacts with the water to form ammonium hydroxide. This contains ammonium (NH_4^+) and hydroxide ions:

ammonia + water → ammonium ions + hydroxide ions
$$NH_3(g) + H_2O(l) \rightarrow NH_4^+(aq) + OH^-(aq)$$

Ammonium hydroxide is a weak alkali, so there are only a few hydroxide ions.

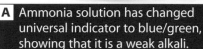

A Ammonia solution has changed universal indicator to blue/green, showing that it is a weak alkali.

1 Which ions are found in ammonia solution?

2 Why is ammonia solution alkaline?

3 How is ammonium hydroxide different to other alkalis like sodium hydroxide and potassium hydroxide? (*Hint*: Look back at Topic C2.32 to help you.)

As ammonia solution is an alkali, it can be neutralised by an acid to form an **ammonium salt**. For example:

ammonium hydroxide + sulfuric acid → ammonium sulfate + water
$$2NH_4OH(aq) + H_2SO_4(aq) \rightarrow (NH_4)_2SO_4(aq) + H_2O(l)$$

You can do this practically by adding sulfuric acid to ammonia solution. The exact amounts can be found by doing a titration.

Ammonium sulfate is an important fertiliser, as it contains the nitrogen essential for the growth of plants. Another ammonium salt used as a fertiliser is ammonium nitrate (NH_4NO_3). This is an even better fertiliser than ammonium sulfate as it contains a larger percentage of nitrogen. However, ammonium nitrate is explosive and has to be handled carefully.

4 Describe how you would make ammonium sulfate by a titration method.

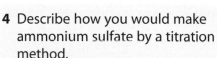

B The crop on the left was grown using nitrogen fertiliser. The crop on the right was grown without nitrogen.

5 a Write a word equation for the preparation of ammonium nitrate.

H

b Write a balanced symbol equation for this reaction. (Nitric acid has the formula HNO_3.)

6 The relative atomic masses of hydrogen, nitrogen, oxygen and sulfur can be found on page 225. Use these to show that the percentage of nitrogen in ammonium nitrate is greater than the percentage of nitrogen in ammonium sulfate.

potassium deficiency

nitrogen deficiency

phosphorus deficiency

C Farmers know which fertiliser to use by looking closely at the plants.

Fertilisers produced chemically are called **inorganic fertilisers**. As well as ammonium nitrate, they also often contain potassium and phosphorus compounds, as these two elements are also essential for healthy plant growth. These are called NPK fertilisers. To make them:

- potassium chloride is crushed
- phosphate rock is reacted with concentrated sulfuric acid to obtain phosphoric acid. This is then reacted with ammonia solution, to form ammonium phosphate.

The compounds are then mixed in the correct proportions and sold as small pellets. Diagram D shows how all the processes are connected.

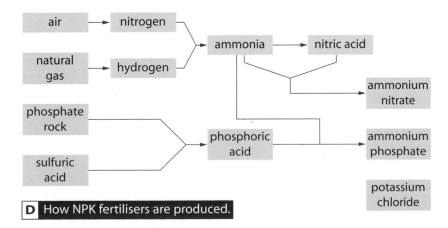

D How NPK fertilisers are produced.

7 Why is the potassium chloride crushed before being mixed into the fertiliser mix?

8 Write a word equation showing the production of ammonium phosphate.

9 Design a leaflet explaining how fertilisers are made and why they are needed.

Investigative Skills Assessment 2

Richard and Nikki were trying to find out the effect of concentration on reaction rate. Limestone contains calcium carbonate, so they decided to react limestone with hydrochloric acid. Hydrochloric acid reacts with the calcium carbonate to form calcium chloride, carbon dioxide and water. They investigated the effect of varying the concentration of hydrochloric acid on the rate of the reaction between calcium carbonate and hydrochloric acid. They used the apparatus shown in diagram A, and measured how long it took to collect 40 cm³ of carbon dioxide.

Their results are shown in table B. Each experiment was carried out three times.

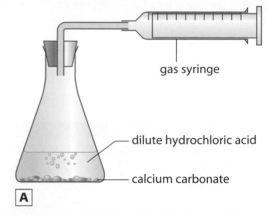

gas syringe

dilute hydrochloric acid

calcium carbonate

A

Concentration of acid (mol/dm³)	Time to collect 40 cm³ of carbon dioxide (seconds)			
	Experiment 1	Experiment 2	Experiment 3	Average time
0.2	251.3	240.9	259.0	
0.4	63.5	64.9	63.8	
0.6	27.7	27.9	28.2	
0.8	15.8	20.5	16.1	
1.0	11.1	10.2	8.9	

B

1 One of the results in the experiment was anomalous. Which one? *(1 mark)*

2 Suggest a reason for this anomalous result. *(1 mark)*

3 How could Richard and Nikki have increased the precision of their results? *(1 mark)*

4 Copy the table. Calculate the average times, excluding the anomalous result. *(2 marks)*

5 a How could Richard and Nikki have improved the validity of their investigation? *(1 mark)*
   b How would the change you have described in part **a** improve the validity? *(1 mark)*

6 What is the range of results for the most dilute acid? Suggest why there is such a range. *(1 mark)*

7 Why do you think that Richard and Nikki decided not to try concentrations greater than 1.0 mol/dm³? *(2 marks)*

8 Michael and Anna decided to investigate this reaction by placing the whole apparatus on a balance, and then found the mass every 10 seconds. They obtained a graph like graph C. They did not have time to do any repeats for the experiment and only used one concentration of acid.
   a Which part of the graph shows where the reaction was fastest. *(1 mark)*
   b Why has the graph flattened out? *(1 mark)*

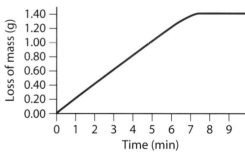

C Graph showing loss of mass against time.

9 ✎ Three other groups used the same method as Michael and Anna, but each group used a different concentration of acid. They also did not have time to do any repeats.
   a How reliable were the results of the four groups who measured mass loss? Explain your answer. *(1 mark)*
   b Each group found the loss of mass after 30 seconds. How could this mass loss be used to compare the effect of concentration on the rate of this reaction? *(2 marks)*
   c What results would you expect? *(1 mark)*

Total = 16 marks

1 Look at the following diagram of an atom.
  a What are the particles labelled A, B and C?
     (*3 marks*)

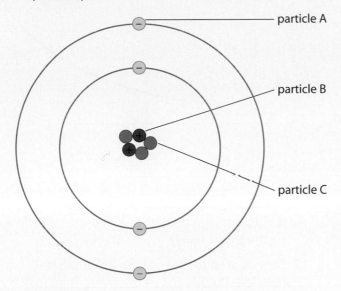

particle A

particle B

particle C

  b What is the mass number of this atom? (*1 mark*)
  c What is the atomic number of this atom?
     (*1 mark*)
  d What is the electronic structure of this atom?
     (*1 mark*)
  e To which group of the Periodic Table does this atom belong? (*1 mark*)

2 Calcium oxide contains calcium ions (Ca^{2+}) and oxide ions (O^{2-}). Copy and complete the diagrams to show the electron structure of these two ions.
  (*2 marks*)

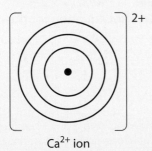

Ca^{2+} ion

2+

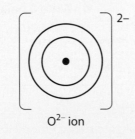

2−

O^{2-} ion

3 Here is a list of substances: H_2O, Ar, CO_2, H_2SO_4, NH_3, Br_2
  a Which are elements? (*1 mark*)
  b Which are compounds? (*2 marks*)

4 a Calculate the relative formula mass of ammonium nitrate, NH_4NO_3. (Relative atomic masses: H = 1, N = 14, O = 16) (*1 mark*)
  b Calculate the percentage by mass of oxygen in ammonium nitrate, NH_4NO_3. (*1 mark*)
  c Calculate the mass of one mole of ammonium nitrate, NH_4NO_3. (*1 mark*)

5 a Paracetamol can be made in different ways. Why do we use the method with the highest atom economy? (*1 mark*)
  b When making paracetamol, less product is made than is expected. Give **two** reasons why.
     (*2 marks*)

6 The diagram shows a molecule of ammonia, NH_3. What do the lines between the atoms represent?
(*1 mark*)

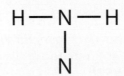

7 Explain why copper is used to make electrical cables. (*1 mark*)

8 a Choose from the words below to complete the following sentence. (*1 mark*)

  molecules, atoms, electrons, ions

  Salt conducts electricity when dissolved in water because its _____ can move.

9 The Haber process is used to make ammonia.
   a Where do the two raw materials used in the Haber process come from? *(2 marks)*
   b Write a word equation for the Haber process. *(1 mark)*
   c What is the ammonia used for? *(1 mark)*
   d What are the conditions of temperature and pressure used in the Haber process? *(2 marks)*
   e Why is an iron catalyst used? *(1 mark)*
   f Why are the unreacted gases recycled? *(1 mark)*

10 Here are some results for the reaction between sodium thiosulfate and hydrochloric acid. The concentration of the sodium thiosulfate has been changed. This is often called 'the disappearing cross' method as the sulfur formed obscures a cross written on a piece of paper.

Grams of sodium thiosulfate in 100 cm³ of solution used	Time for cross to be obscured (s)
7.5	21
6.0	28
4.5	33
3.0	54
1.5	122

   a What do these results show? *(1 mark **HSW**)*
   b Why is it difficult to get accurate results using this method? *(2 marks **HSW**)*
   c The data can be shown on a line graph. Sketch the graph that you would expect to see. *(2 marks **HSW**)*
   d What variables must be controlled during this experiment? *(2 marks **HSW**)*

11 Manganese dioxide is the catalyst used in the breakdown of hydrogen peroxide to water and oxygen.
$$2H_2O_2(aq) \rightarrow 2H_2O(l) + O_2(g)$$
   a How could you collect the oxygen produced by this reaction? *(1 mark)*
   b How would the rate of this reaction change if powdered manganese dioxide was used instead of a lump? Explain your answer. *(2 marks)*

c This is a graph showing what happens when the reaction mixture is heated.

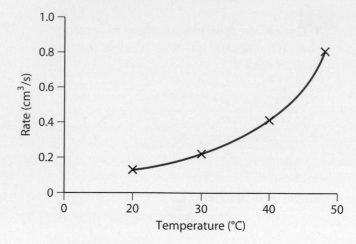

How does increasing the temperature affect the rate of this reaction? *(1 mark)*
   d Explain your answer to **c** in terms of collision theory. *(1 mark)*

12 Tom wanted to make some insoluble lead chloride ($PbCl_2$).
   a What is the name of the reaction in which two solutions react to form an insoluble salt? *(1 mark)*
   b Tom chose lead nitrate and potassium chloride for this reaction. Write a word equation for this reaction. *(1 mark)*

13 Ammonia dissolves in water to form ammonium hydroxide (NH_4OH). This can be reacted with hydrochloric acid (HCl) to form ammonium chloride (NH_4Cl).
   a What is the name for a reaction where an acid reacts with an alkali? *(1 mark)*
   b Write a balanced symbol equation for this reaction. *(1 mark)*

Assessment exercises Higher

1 Copy and complete the table to show the structure of some atoms. *(3 marks)*

Atom	$^{23}_{11}$Na	K	F
Atomic number		19	
Mass number		39	
Number of protons			9
Number of neutrons			10
Number of electrons			
Electronic structure			

2 Calcium ions have a charge of 2+. They contain 20 protons. How many electrons do they contain? *(1 mark)*

3 Which type of structure do the substances in the following table have? Choose from: ionic, simple molecular, giant covalent, metallic. *(4 marks)*

Substance		A	B	C	D
Melting point (°C)		612	1713	-35	637
Boiling point (°C)		1845	2230	47	983
Electrical conductivity as	solid	conducts	does not conduct	does not conduct	does not conduct
	liquid	conducts	does not conduct	does not conduct	conducts
	solution (aq)	insoluble	insoluble	does not conduct	conducts

4 Ammonia is a simple molecular substance. It has the formula NH_3.
Copy and complete the diagram to show the arrangement of electrons in ammonia. *(1 mark)*

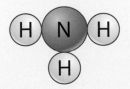

5 The electrolysis of sodium chloride solution is a major industrial process. The three products are chlorine gas, hydrogen gas and sodium chloride solution.

  a Identify gas A and gas B in the diagram. *(1 mark)*

  b The half equation for the formation of hydrogen is:
  $$2H^+ + 2e^- \rightarrow H_2$$
  Is this an oxidation or a reduction process? Explain your answer. *(1 mark)*

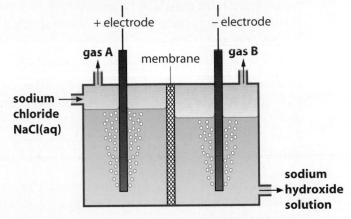

6 Sodium reacts with oxygen to make sodium oxide.
$$4Na(s) + O_2(g) \rightarrow 2Na_2O(s)$$

  a What mass of sodium oxide can be made from 115 g of sodium? (Relative atomic masses: O = 16, Na = 23) *(3 marks)*

  b The reaction was done three times to measure how much sodium oxide was made. The three results were 137 g, 142 g and 141 g.

   i Calculate the average mass of sodium oxide formed. *(1 mark)*

   ii Calculate the percentage yield. *(1 mark)*

   iii Explain why the experiment was done three times. *(1 mark **HSW**)*

  c Give **two** reasons why reactions do not give 100% yields. *(2 marks)*

7 Sodium metal is made by the electrolysis of molten sodium chloride.
$$2NaCl \rightarrow 2Na + Cl_2$$

  a Calculate the atom economy for this process. (Relative atomic masses: Na = 23, Cl = 35.5) *(1 mark)*

  b Explain why reactions with a high atom economy are better than those with a low atom economy. *(1 mark)*

8 N_2O_4 molecules are pale yellow, while NO_2 molecules are dark brown. At room temperature they are both present in an equilibrium mixture. The equation is:

$$N_2O_4(g) \text{ (+ heat)} \rightleftharpoons 2NO_2(g)$$

  a What does $\rightleftharpoons$ mean? *(1 mark)*

  b Is the forward reaction endothermic or exothermic? *(1 mark)*

  c What colour would you expect the equilibrium mixture to be? *(1 mark)*

  d What will happen to the equilibrium if it is heated? *(1 mark)*

  e What will happen to the equilibrium if the pressure is increased? *(1 mark)*

  f What will happen to the rate of the reaction if the pressure is increased? *(1 mark)*

9 Here are some results for the reaction between sodium thiosulfate and hydrochloric acid. The concentration of the sodium thiosulfate has been changed. This is often called 'the disappearing cross' method as the sulfur formed obscures a cross written on a piece of paper.

Concentration of sodium thiosulfate solution (mol/dm³)	Time for cross to be obscured (s)
0.30	21
0.24	28
0.18	33
0.12	54
0.06	122

  a Describe what these results show. (*1 mark* **HSW**)

  b Why is it difficult to get accurate results using this method? (*1 mark* **HSW**)

  c How could you increase the reliability of these results? (*1 mark* **HSW**)

  d The data can be shown on a line graph. Sketch the graph that you would expect to see. (*2 marks* **HSW**)

  e Which variables must be controlled? (*2 marks* **HSW**)

  f Explain what the units for concentration mean. (*2 marks*)

  g If 100 cm³ of sodium thiosulfate of concentration 0.30 mol/dm³ are used, how many moles of sodium thiosulfate are there in this volume of solution? *(2 marks)*

10 Tom wanted to make some insoluble lead chloride ($PbCl_2$).

  a What is the name of the reaction in which two solutions react to form an insoluble salt? *(1 mark)*

  b Copy and complete this balanced symbol equation:
$$Pb(NO_3)_2(aq) + 2KCl(aq) \longrightarrow \underline{\quad}(\underline{\,}) + \underline{\quad}(\underline{\,})$$
(2 marks)

11 Ammonia dissolves in water to form ammonium hydroxide (NH_4OH). This can be reacted with hydrochloric acid to form an ammonium salt.

  a Ammonium hydroxide is said to be a weak alkali. What does this mean? *(1 mark)*

  b Write a balanced symbol equation for this reaction. *(1 mark)*

12 Name **two** elements, other than nitrogen, that plants need for healthy growth. *(2 marks)*

Glossary

acid Chemical that can form hydrogen ions in water to give a solution of pH less than 7.

activation energy Minimum energy that particles must have to react.

alkali Base that dissolves in water to form hydroxide ions and give a solution of pH greater than 7.

alkali metals Elements in Group 1 of the Periodic Table.

ammonia NH_3, a compound used to make fertilisers.

ammonium salt Salt formed when ammonia solution reacts with an acid.

anhydrous Crystals that do not contain any water.

atom Particle with no electric charge containing protons, neutrons and electrons. The smallest part of a chemical element.

atom economy (atom utilisation) The percentage of the mass of the products of the reaction that is the desired product.

atomic number Number of protons in the nucleus of an atom.

H backwards reaction The reverse of a chemical reaction as shown by the symbol equation.

base (chemistry) Substance that reacts with an acid.

burette Apparatus used to measure volumes of liquids.

catalyst Substance that speeds up a chemical reaction, but is not itself used up.

collision theory Theory that all particles must collide with enough energy to react.

combustion Burning a substance.

compound Substance made from different elements chemically joined together.

concentrated Large amount of solute dissolved in a solvent such as water.

concentration Measure of how much solute is dissolved in a solvent such as water.

covalent bond A pair of electrons shared between two atoms.

crystallisation Formation of crystals from a solution.

dilute Small amount of solute dissolved in a solvent such as water.

H dynamic equilibrium Equilibrium where the forward reaction and backwards reaction are both occurring at the same time and at the same rate.

electrolysis Decomposition of an ionic compound using electricity.

electron A particle inside atoms with a negative electric charge.

electrostatic attraction The attraction between positive and negative electric charge.

element Substance containing only one type of atom.

H empirical formula Shows the simplest ratio of atoms of each element in a substance.

endothermic Chemical reaction that takes in heat energy.

end point The point in a titration where the acid and alkali have exactly reacted together.

energy level Orbit in which electrons move around the nucleus inside atoms (also called a shell).

H equilibrium Chemical reaction in which the forward and backwards reactions occur at the same time.

exothermic Chemical reaction that gives out heat energy.

H forward reaction Chemical reaction as shown by the symbol equation.

giant covalent structure Structure containing billions of atoms in a network linked together by covalent bonds (also called macromolecular).

giant structure Regular, continuous structure of atoms or ions.

Haber process Industrial process used to make ammonia.

H half equation Balanced equation showing what happens at an electrode during electrolysis.

halogens Elements in Group 7 of the Periodic Table.

hydrated Crystals that contain water.

hydrogen ions H^+ ions are formed when an acid dissolves in water.

hydroxide ions OH^- ions found in all alkaline solutions.

indicator Chemical that changes colour at the end point.

inorganic fertilisers Manufactured chemicals containing nitrogen, which are added to the soil to increase crop yield.

insoluble Substance that does not dissolve in a liquid such as water.

H intermolecular forces Weak attractive forces between molecules.

ion Electrically charged particle containing a different number of protons and electrons.

ionic bonding Attraction between positive and negative ions.

ionic compound Substance made from positive and negative ions.

ionic equation Equation that shows what happens to the ions in a reaction.

ionic structure Substance made up of positive and negative ions.

isotope Atoms of the same element with the same number of protons but a different number of neutrons (the same atomic number but a different mass number).

lattice A regular, continuous structure of atoms or ions.

macromolecular structure Giant covalent structure.

mass number Number of protons plus the number of neutrons in an atom.

metallic Substance that is a metal.

H metallic bonding Attraction between positive metal ions and delocalised electrons.

H molar volume Volume of 1 mole of gas at room temperature and pressure. It is 24 dm³.

mole 602 204 500 000 000 000 000 000 of anything.

molecular formula Shows the number of atoms of each element in one molecule.

molecule Particle made of atoms joined to each other by covalent bonds.

monatomic Substance made up of lots of individual atoms.

nanomaterials Substances made of particles that are between 1 and 100 nm in size.

nanoparticles Particles that are between 1 and 100 nm in size.

nanoscience The study of structures that are between 1 and 100 nm in size.

neutralisation Reaction of an acid with a base.

neutron A particle inside the nucleus of an atom with no electric charge.

noble gases Elements in Group 0 of the Periodic Table.

nucleus (chemistry) The centre of an atom containing protons and neutrons.

oxidation Gain of oxygen (in terms of oxygen); loss of electrons (in terms of electrons).

percentage yield Mass of product obtained from a reaction expressed as a percentage of the theoretical maximum mass.

pH scale Scale used to measure the strength of acids and alkalis.

pipette Apparatus used to measure exact volumes of liquid.

precipitation reaction Reaction in which two solutions mix and form an insoluble salt.

products Substances formed during a chemical reaction.

proton A particle with a positive charge found in the nucleus of an atom.

proton number Number of protons in an atom (atomic number).

rate of reaction Speed of a reaction, expressed as the amount of reactant used up or amount of product formed over time.

reactants Substances that react together in a chemical reaction.

reduction Loss of oxygen (in terms of oxygen); gain of electrons (in terms of electrons).

H relative atomic mass (A$_r$) Average mass of an atom relative to $^1/_{12}$ the mass of a ^{12}C atom.

H relative formula mass (M$_r$) Sum of all the relative atomic masses of the atoms in a formula; the average mass of all the atoms in a formula relative to $^1/_{12}$ the mass of a ^{12}C atom.

H relative mass Mass of a particle relative to $^1/_{12}$ the mass of a ^{12}C atom.

reversible reaction Chemical reaction that can go in either direction.

shell (chemistry) Orbit in which electrons move around the nucleus inside atoms (also called an energy level).

simple molecular structure Substance made up of lots of separate molecules.

smart materials Materials which have one or more properties that change in different conditions.

soluble Substance that dissolves in a liquid such as water.

solute Solid that dissolves in a solvent to make a solution.

solution Solute dissolved in a liquid such as water.

specific (about a catalyst) The fact that most catalysts work for one reaction only.

strong acid Acid which forms lots of hydrogen ions when added to water.

strong alkali Alkali which forms lots of hydroxide ions when added to water.

thermal decomposition Splitting up a compound by the action of heat.

titration Method of finding volumes of solutions that just react together.

weak acid Acid that does not form many H^+ ions in water.

weak alkali Alkali that does not form many OH^- ions in water.

H yield (chemistry) Mass of a substance made in a chemical reaction.

The physics of sailing

A Sailing a yacht involves a lot of forces and plenty of hard work.

B Nuclear power can be dangerous, but also useful.

Ocean motion

Imagine trying to harness the energy of the wind to sail a large yacht at high speeds. And in the right direction! How would you be able to work out how fast you were going and where you had ended up? How could you use the wind to continue on the right course to win a race? Sailing involves a lot of science. Navigation and the design of the boats are based on a lot of scientific ideas. Boats can travel faster and more safely than ever before as a result of new technologies.

Nuclear power

Nuclear reactions produce a lot of energy. They can produce enough energy to power a submarine, or light up stars.

By the end of this unit you should be able to:

* describe the way things move, how they speed up or slow down, and what effect this has on their movement energy
* explain what momentum is and how to calculate it
* describe what happens to radioactive substances when they decay, including the nuclear reactions called fission and fusion.

1 Write down five activities sailors have to carry out when sailing.

2 List as many scientific ideas as you can think of that might be useful in completing a modern ocean-crossing yacht race.

3 List as many words as you can think of that are connected with nuclear power.

Speed, distance and time

By the end of this topic you should be able to:

- describe what a distance–time graph shows
- construct distance–time graphs
- **H** calculate speed using a distance–time graph

Sailors have to carefully record their boat's movement in order to know where they are. By noting down their position over time, a **distance–time graph** for the boat can be drawn. This will also show how far it travels in a certain time (the boat's **speed**).

By plotting the values in the sailors' data table, we can get a picture of how the boat moved. To plot a distance–time (d–t) graph, you must always put time along the x-axis. At each point in time, you can then plot the distance the object has moved by that time.

A How fast is this yacht moving?

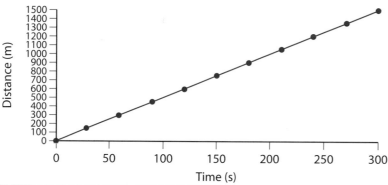

C Distance–time graph of a yacht race.

Time (s)	Distance (m)
0	0
30	150
60	300
90	450
120	600
150	750
180	900
210	1050
240	1200
270	1350
300	1500

B Distance and time data of a yacht race.

Speed is shown on a distance–time graph by looking at how steep the slope is. If it is very steep, then the object has moved a large distance in a short time. This means it is moving fast.

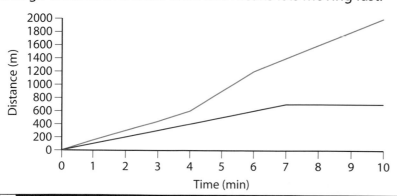

D Distance–time graphs for two boys walking. Who is moving the fastest?

As the speed changes, the angle of the slope changes. For example, in graph D, the boy shown by the blue line moves faster from 4 to 6 minutes and then slows down again. The boy with the red line stops when he reaches 700 metres. You can see this because his line goes flat.

1 Write down three pieces of information you can get from a distance–time graph.

2 Which boy is moving slower in graph D?

The speed can be calculated from a distance–time graph. Speed is calculated using the equation:

$$\text{speed} = \frac{\text{distance}}{\text{time}}$$

$$v = \frac{d}{t}$$

$$[\text{metre/second, m/s}] = \frac{[\text{metre, m}]}{[\text{second, s}]}$$

This means that calculating the gradient of the line on a distance–time graph would give us the speed. For example, in graph B, the yacht's speed can be found by looking at the points where the line starts and finishes. It moves a distance of 1500 metres in 300 seconds:

Example

Calculate the speed of a competitor who completes 1500 m in 300 s.

$$v = \frac{d}{t}$$
$$= \frac{1500}{300}$$
$$= 5 \text{ m/s}$$

3 Calculate the start speed of each boy in graph D.

Distance–time graphs always show the distance increasing without showing which way the object moves. This means the actual speed of movement can always be found.

4 Geraldine went hiking with three friends. She had the route written down on a card (see table E).
  a Work out the 'total distance' entries on the card that have been left out.
  b Draw a distance–time graph for Geraldine's hiking trip.

Time (min from start)	Direction	Distance to walk (m)	Type of ground	Landmarks to help navigate	Total distance (m)
Start	east	100	country road	look for stile	100
5	north-east	400	path across fields	head up to sheep pen	500
20	north-east	500	ridge	rock pile	
40	rest for lunch	0			
60	south	600	rocky path	go down steep hill path	
80	south-west	400	disused railway	to car park	
100	arrive at car park				

E

5 a Draw an example distance–time graph to show your journey to school.
  b Include labels on the graph to show places where you were:
    (i) not moving
    (ii) moving fastest.

  c The speed can be calculated from a d-t graph by calculating the gradient (slope) of the line. Show example calculations of the gradient to find out how fast you were travelling in each section of the graph.

Velocity

By the end of this topic you should be able to:

- explain what the term velocity means
- describe what a velocity–time graph shows
- construct a velocity–time graph
- **H** • calculate the distance travelled using a velocity–time graph.

Velocity is a measure of speed in a certain direction.

1 What is the difference between speed and velocity?

As well as distance–time graphs, we can also draw
velocity–time (v–t) graphs, to show what happens during a
journey. Velocity–time graphs tell us directly how fast the object
is going, but also if that movement is forwards or backwards.

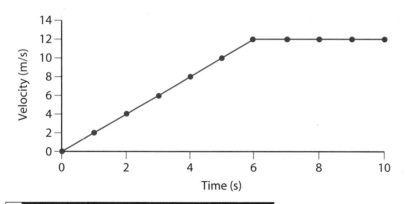

B Velocity–time graph of a girl running 100 m.

A velocity–time graph always has time along the *x*-axis, and
velocity on the *y*-axis. At each moment in time, the velocity is
plotted. In order to show an object moving backwards, the
velocity would be plotted as a negative number, below the *x*-axis.

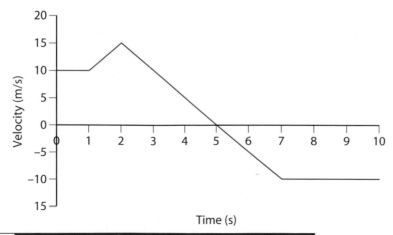

C Velocity–time graph for a car travelling in a car park.

A Which boat is moving fastest? But
which is moving fastest towards the
finishing line?

2 a In graph B, what is the highest
velocity the girl reaches?
  b What was her velocity after
4 seconds?

If we plot a velocity–time graph for a car moving in a car park, it might look like the one in graph C. At the start, the car is moving forwards. After moving a bit faster, it then slows down and stops (v = 0 m/s) after 5 seconds. Then it moves backwards, in the opposite direction from the way it was moving at the start. You can see this on the graph where the line goes below the x-axis down to −10 m/s. Finally it is reversing at 10 m/s.

If an object is moving more quickly, the line on its velocity–time graph will be higher up. If you look at the area between the line and the x-axis, you can see how far the object will have moved. More area under the line means more distance travelled.

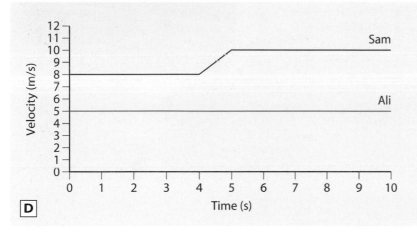

D

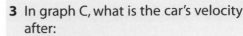

3 In graph C, what is the car's velocity after:
  a 1 second
  b 5 seconds
  c 6 seconds?

4 How can you quickly see from graph C that car goes backwards for a time?

5 Graph D shows the velocities of two boys, Sam and Ali, while cycling, over a 10 second period.
  a Which boy is moving slower?
  b Which boy changes his speed?
  c Which boy moves further in the 10 seconds shown?
  d Explain how you worked out your answer to **c**.

H To find the exact distance an object has travelled, you need to calculate the area under the line on the velocity–time graph.

Example

Calculate the distance run by a dog across a field as shown in graph E.

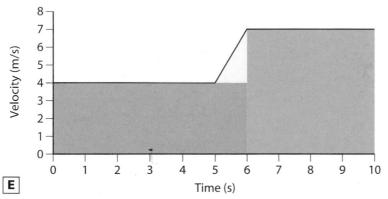

E

Distances covered:
red area: 4 m/s × 6 s = 24 m
blue area: 7 m/s × 4 s = 28 m
yellow area: $\frac{1}{2}$ × 3 m/s × 1 s = 1.5 m
total area: 24 m + 28 m + 1.5 m = 53.5 m
So the dog ran 53.5 metres across the field.

6 **a** Calculate how far each boy went in the time shown in graph D.
  b What was the difference in the distances they went?

7 **a** The speedometers in lorries are fitted with a machine called a tachograph. Tachographs automatically produce a velocity–time graph of the lorry's movements. Write a paragraph which explains to learner lorry drivers what the velocity–time graph can show. Include an example graph in your answer.
  H **b** Label the graph to show how the exact information can be worked out using the graph.

Acceleration

P
H

By the end of this topic you should be able to:

- calculate acceleration by using an equation
- describe how acceleration is shown on a velocity–time graph
- calculate acceleration from a velocity–time graph.

When the velocity of an object changes, we say that it is accelerating. This can be seen on a velocity–time graph.

To calculate **acceleration**, you must work out how much an object's velocity has changed. You also need to know how long it took to make that change in velocity. This information can then be used in this calculation:

A This car is accelerating.

$$\text{acceleration} = \frac{\text{change in velocity}}{\text{time taken for change}}$$

$$a = \frac{v_2 - v_1}{t_2 - t_1}$$

$$[\text{metre/(second)}^2, \text{m/s}^2] = \frac{[\text{metre/second, m/s}]}{[\text{second, s}]}$$

Example

Calculate the acceleration of the car in graph B.
From 0 seconds to 3 seconds, the car accelerates from a velocity of 0 m/s to 24 m/s.

$$a = \frac{24 - 0}{3 - 0} = \frac{24}{3} = 8 \text{ m/s}^2$$

acceleration = 8 m/s²

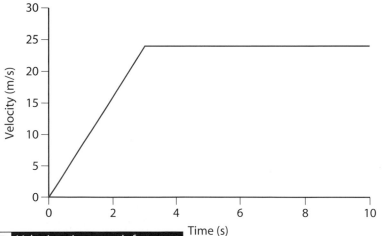

B Velocity–time graph for a car.

Even if it gets slower, the object is still said to accelerate. The figure given for the acceleration is negative, to show that the object's velocity is reducing.

1 At the start of a horse race, a jockey and his horse accelerate from standing still to a velocity of 10 m/s in 2 seconds. What is the horse's acceleration?

2 A train slows down as it approaches a station. If its velocity reduces from 15 m/s to 5 m/s in 4 seconds, what is the train's acceleration?

On a velocity–time graph, the slope of the line shows us the acceleration. The steeper the line, the greater the acceleration. If the line is less steep, the acceleration is less.

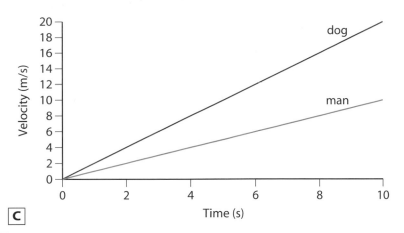

C

P

3 In graph C, does the man or the dog have the greater acceleration?

H

By calculating the slope of the line on a velocity–time graph, we are calculating the acceleration. The calculation to find the slope of the line is the same as that to find the acceleration:

$$\text{slope} = \frac{\text{change in velocity}}{\text{time for that change}}$$

4 a From graph C, calculate the slope of each line.
b What is the acceleration of:
   (i) the man
   (ii) the dog?

P

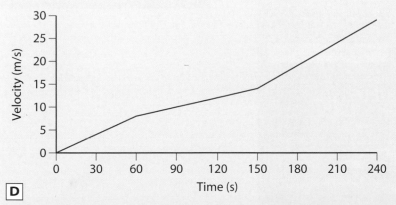

D

P

5 Look at the velocity–time graph D for a boat in a sprint race.
  a Describe how the velocity of the boat changes over the time shown on the graph.
  b How does the acceleration of the boat change?
H
  c Calculate the acceleration in the following intervals:
    (i) 0–30 s
    (ii) 60–150 s
    (iii) 150–180 s
    (iv) 180–240 s

6 In a sales leaflet, a motorised scooter is advertised as having an acceleration of 6 m/s².
  a Write a technical note to go at the end of the leaflet to explain what acceleration is and what '6 m/s²' means.
  b Include a velocity–time graph to support your technical description.

D
P
P

Adding forces

By the end of this topic you should be able to:

- add forces acting in different directions
- describe what a resultant force is.

Forces are measured in **newtons (N)**. The direction in which they act is always given. For example, your weight might be 560 N, but you would have to say it is 560 N downwards. If two forces of the same size act in opposite directions on an object, they are said to be **balanced**. For example, if a boy hangs on a tree branch, the force of the branch pulling up on his arms is equal to the boy's weight. The forces up and down are balanced.

1 What does it mean to say that the forces are balanced?

2 If a strongman lifts a weight of 280 N and the forces on it are balanced, how much force must he be pushing up with?

If the forces acting in opposite directions are different, they are said to be **unbalanced**. This would happen to a sailing boat if the wind filled its sails and pushed it forward more than the drag forces pushing it back.

branch force = 560 N

weight = 560 N

A The forces on this boy are balanced.

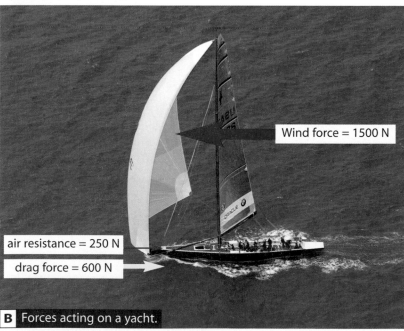

Wind force = 1500 N

air resistance = 250 N

drag force = 600 N

B Forces acting on a yacht.

3 A stone dropped from a cliff has unbalanced forces acting on it. Explain what this means.

4 Give an example of when a car has:
   a balanced forces acting on it
   b unbalanced forces acting on it.

If there is more than one force in one line, they can be added together to give a total force. This total, or overall, force is called the **resultant force**. A sailing boat being pushed forward by the wind would have a force slowing it down caused by drag with the water, and one caused by air resistance. These two forces which slow the boat down could be added together before comparing them with the forward force. The resultant force on the yacht in photograph B is:

1500 − (600 + 250) = 650 N forwards.

C

5 What is the resultant force on the doll in diagram C?

6 Draw a picture of a car.
   a Add force arrows to your picture for the forward force from the engine, friction and air resistance.
   b Label the forces with values so that the resultant force on the car is 370 N forwards.

Forces can be added up separately in any direction. This means that vertical forces can be added together to find the vertical resultant. Separately, horizontal forces can be added to find the horizontal resultant force.

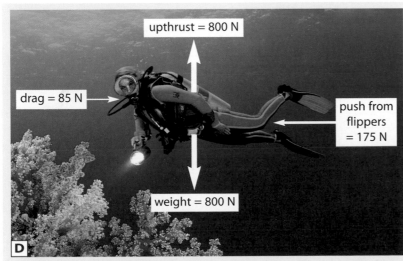

D

7 Thinking about horizontal and vertical forces separately, what is the resultant force on the diver in diagram C?

8 a Draw a picture of a plane with four forces acting on it: weight, lift, thrust and drag.
   b Explain how big each force might need to be so that it could have a resultant force which acts:
   (i) upwards
   (ii) forwards
   (iii) downwards
   (iv) backwards.

167

Forces and movement

By the end of this topic you should be able to:

- describe the effects of resultant forces on stationary objects
- describe the effects of resultant forces on moving objects
- use the equation which relates resultant force, mass and acceleration.

An object with a resultant force acting on it will change its movement. It may change direction, it may accelerate, or it may slow down.

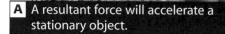
A A resultant force will accelerate a stationary object.

B Stationary objects with balanced forces stay stationary.

If there is no resultant force, all forces are balanced and the movement of the object will remain the same. If it is not moving, and has only balanced forces, then it will remain stationary.

Starting movement	Forces balanced or unbalanced?	Resultant force?	Effect on movement
Not moving	balanced	no	remains still
Not moving	unbalanced	yes	starts moving in the direction of the resultant **a**
Moving	balanced	**b**	carries on in the same direction at the same speed
Moving	**c**	yes	changes speed or direction, or both

C

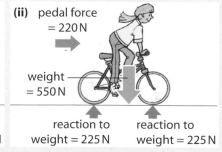

(i) weight = 550 N / reaction to weight = 225 N / reaction to weight = 225 N

(ii) pedal force = 220 N / weight = 550 N / reaction to weight = 225 N / reaction to weight = 225 N

D

1 Copy and complete table C.

2 What is needed to cause acceleration?

3 **a** Say what would happen to the movement of the girl on the bike in parts **(i)** and **(ii)** of diagram D.
   b Explain your answers.

Sir Isaac Newton

The famous English scientist Sir Isaac Newton worked out a lot of the science about forces and movement. He worked out that the acceleration depends on the overall, or resultant, force and the **mass** of the object. He wrote this as an equation:

resultant force = mass × acceleration
$$F = m \times a$$
[newton, N] = [kilogram, kg] × [metre/(second)^2, m/s^2]

Example

A sailing boat has a mass of 1200 kg. If the boat accelerates at 4 m/s^2, calculate the resultant force acting on it.

$$F = m \times a$$
$$= 1200 \times 4 = 4800 \text{ N (forwards)}$$

This resultant force of 4800 N forwards is the total of the wind pushing forwards minus the drag and air resistance pushing backwards.

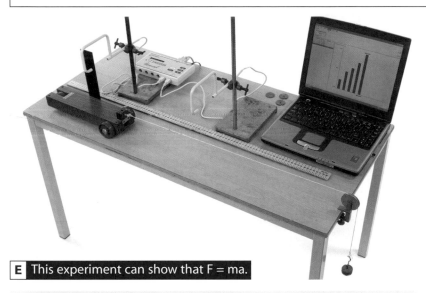

E This experiment can show that F = ma.

4 **a** A stone dropped from a cliff has a mass of 0.9 kg. It accelerates down at 10 m/s^2. Calculate the force acting on the stone.
b What do we call this force accelerating the stone?

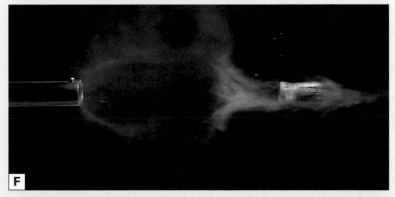

F

5 A bullet is fired and accelerates along the barrel of the gun at 2000 m/s^2 (see photograph F). If the force of the explosion pushing the bullet is 10 N, what is the mass of the bullet?

6 A car with a mass of 1000 kg has a forward force caused by the engine of 700 N. The friction on it is 150 N and the air resistance is 50 N.
a What is the resultant force on the car?
b Calculate the acceleration of the car.

7 Imagine you are Sir Isaac Newton. Write a letter home to your mother in which you explain very simply the science you have discovered about forces and motion. Include diagrams to show four situations: a stationary object with balanced forces, and one with unbalanced forces acting on it, a moving object with balanced forces, and one with unbalanced forces. Also explain the equation that relates force, mass and acceleration.

Weight and terminal velocity

By the end of this topic you should be able to:

- calculate an object's weight
- describe what terminal velocity is and how it comes about
- draw and interpret velocity–time graphs for objects that reach terminal velocity.

All objects are pulled towards the Earth by the force of **gravity**. The amount gravity pulls on a mass is called its **weight**.

> weight = mass × gravitational field strength
> w = m × g
> [newton, N] = [kilogram, kg] × [newton/kilogram, N/kg]

If an object has more mass, then it weighs more. If the **gravitational field** acting on the object is stronger, then it weighs more. On Earth, the strength of gravity is 10 newtons per kilogram ($g = 10$ N/kg).

The weight of an object will make it fall if nothing holds it up. The movement causes air resistance on the object. This drag force slows the object down. Air resistance, or drag, is a frictional force. If an object moves faster through the air (or any **fluid**) the frictional force on the object from the fluid will be greater.

drag = 30 N

drag = 270 N

B Moving faster through a fluid increases the frictional force.

1 What two values are multiplied to find the weight of an object?

2 What is the weight on Earth of:
  a a 1 kg bag of sugar
  b a 55 kg schoolgirl
  c a 1150 kg car?

A How fast can this seed fall?

3 Two boys are racing on their bikes. If Frankie moves at 6 m/s and Mustapha moves at 8 m/s, which boy will feel a stronger force of air resistance on his face?

When an object falls downwards, it accelerates because its weight pulls it down. As it gets faster, the frictional force on it will increase. Since the weight stays the same, this makes the resultant downward force less, and so the object accelerates less. Eventually it will reach a velocity where the frictional force equals the object's weight. At this point the forces are balanced, so the object continues to fall at the same velocity. This is called its **terminal velocity**.

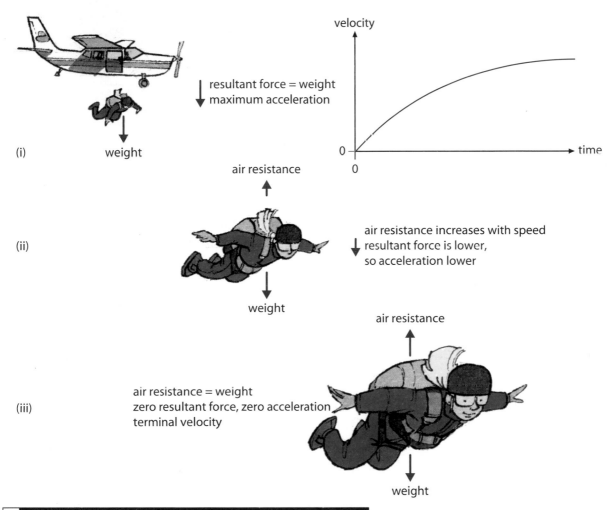

(i)   weight

resultant force = weight
maximum acceleration

(ii)

air resistance

air resistance increases with speed
resultant force is lower,
so acceleration lower

weight

(iii)

air resistance = weight
zero resultant force, zero acceleration
terminal velocity

air resistance

weight

C   Anything falling through a fluid will reach a terminal velocity.

4 Look at diagram C. Write a description of the skydiver's journey from jumping out of the plane until she reaches terminal velocity. Include descriptions of:
   a her speed
   b the forces acting on her.

5 What is the resultant force on a stone which is dropping through water at its terminal velocity?

6 Explain how a boat with a fixed forward force from the wind of 2850 N will reach a constant maximum velocity.

7 On a theme park ride called 'Freefall', your seat is released from a great height to drop vertically. Write an entry to describe this ride for a book called *The Science of Theme Parks*. Your entry should use diagrams showing forces to explain:
   a your velocity
   b your acceleration
   c how you will reach a terminal velocity.

Stopping distances

By the end of this topic you should be able to:

- describe the forces acting on a vehicle travelling at a steady speed
- describe what is meant by the stopping distance of a vehicle
- list what can affect the stopping distance of a vehicle.

1 a What frictional forces act on the car in poster A?
b What causes the car's driving force?

When a vehicle drives at a steady speed, the frictional forces acting on it balance the driving force. For the vehicle to slow down and stop, a resultant backward force needs to be set up. Usually, vehicles such as cars will do this by removing the driving force and increasing the frictional forces from the brakes.

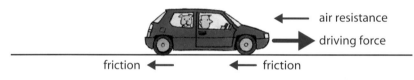

steady speed – no resultant force

air resistance
driving force
friction ← ← friction

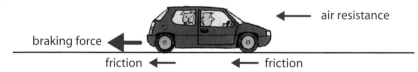

resultant force backwards – car slows down

braking force
air resistance
friction ← ← friction

B Stopping a vehicle.

At 5mph over the 30mph limit, how much further does it take to stop?

THINK!
Slow down

A How quickly can cars stop?

2 What happens to the speed of a car if the forces on it are balanced?

The distance that a car moves while it is stopping is called its **stopping distance**. This is made up of two parts: the **thinking distance** and the **braking distance**.

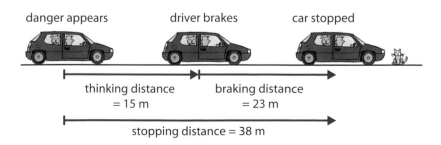

danger appears   driver brakes   car stopped

thinking distance = 15 m
braking distance = 23 m
stopping distance = 38 m

C Stopping distance = thinking distance + braking distance.

The thinking distance is how far a vehicle moves during the driver's **reaction time**. The reaction time, and so the thinking distance, can be made longer by tiredness, drugs or alcohol.

3 Distance travelled = speed × time.
  a How far will a car travel during a driver's reaction time of 0.5 s if the car is going at 17 m/s?
  b If the driver is tired and his reaction time increases to 0.8 s, what will be the new thinking distance?
  c Another driver has drunk a glass of wine and her reaction time is now 1.1 s. She also drives at 17 m/s. How much further does her car go during this reaction time than it would in her normal reaction time of 0.5 s?

When a driver presses the brakes to stop their car, the brakes add a frictional force to slow the car down to a stop. The distance the car takes to stop once the brakes are on will depend on:
• the road conditions
• how good the brakes are
• the speed of the car.

If the road conditions provide less friction, for example, if it is wet or icy, then the braking distance will be longer. If the brakes are old and worn, the frictional force will be less than when they are newer, and the braking distance will also be longer.

If the car is going faster, it will take more time for the brakes to stop it, and so the braking distance will again be longer. If you have to stop the car in a certain distance, the faster the car is going the greater the braking force needed.

4 Explain three situations when the braking distance of a car may increase.

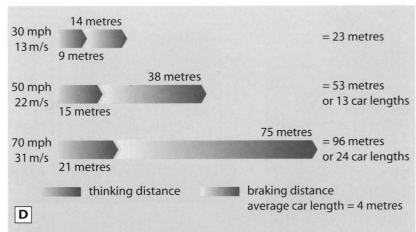

30 mph
13 m/s
14 metres
9 metres
= 23 metres

50 mph
22 m/s
38 metres
15 metres
= 53 metres
or 13 car lengths

70 mph
31 m/s
75 metres
21 metres
= 96 metres
or 24 car lengths

thinking distance
braking distance
average car length = 4 metres

D

5 Diagram D is from the government's website about the Highway Code and shows car stopping distances under normal conditions. You are to write the text that goes with this diagram for an update of the website. Write a paragraph that could be used to explain to the public:
  a what 'stopping distance' means
  b what can affect stopping distance.

Work

By the end of this topic you should be able to:

- describe what is meant by 'work done'
- calculate the work done
- describe what happens when work is done against frictional forces.

Sailing a boat can be hard work. There are often times when ropes have to be pulled with a lot of force to move the sails and the boom. In science, doing **work** means the same as transferring energy.

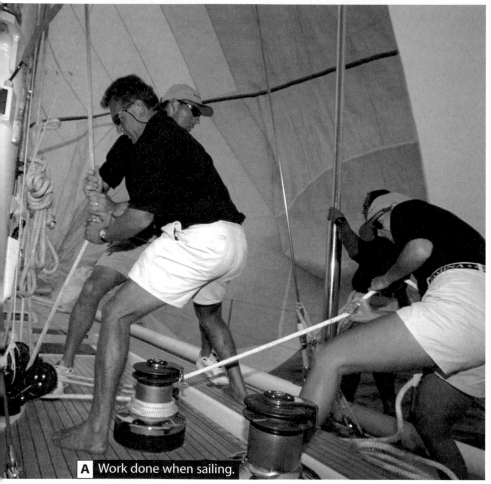

A Work done when sailing.

B Work done = energy transferred.

When energy is changed from one form to another, we say that work is being done. Kicking a ball up, so that its kinetic energy is changed into gravitational potential energy, is doing work. Energy changes can happen in many different ways and may or may not be easy to see happening.

Work is an amount of energy, so it is measured in joules (J). The amount of work done is equal to the amount of energy transferred from one form to another.

1 Give the scientific meaning for 'work done'.

When a force moves an object, energy is transferred and so work is done. The amount of work done can be calculated from the equation:

work done =   force   × distance moved (in the direction of the force)
    W   =   F   ×   d
 [joule, J] = [newton, N) ×   [metre, m]

Example

A sailor lifts a 300 N sail from the deck to a height 4 m higher. How much work does the sailor do?

work done = force × distance moved (in the direction of the force)
    W   = F × d
           = 300 × 4
           = 1200 J

2 Calculate the work done when:
  a a weightlifter lifts a 280 N barbell 1.5 m straight up
  b a boy cycles 760 m against friction forces of 140 N
  c a crane raises a box with a mass of 120 kg upwards 4 m.

3 How much work is done when:
  a a toaster changes 2000 J of electrical energy into heat energy
  b a washing machine changes 5000 J of electrical energy into 3500 J of heat energy, 1200 J of kinetic energy and 300 J of sound energy
  c a car changes chemical potential energy in the petrol into 8.5 kJ of heat energy, 4.2 kJ of movement, 0.4 kJ of sound and 1.1 kJ of light energy?

When work is being done against frictional forces, most of the energy is transformed into heat energy.

C The work done to stop this car has changed kinetic energy into heat energy.

4 Explain why you would expect heat to be produced in the following situations:
  a a plane accelerating along a runway
  b a sailor pulling a rope over a bar very quickly
  c a carpenter sanding a wooden door with sandpaper.

Car stopping distance

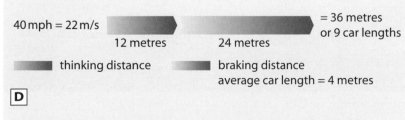

40 mph = 22 m/s

12 metres   24 metres   = 36 metres or 9 car lengths

thinking distance   braking distance
average car length = 4 metres

D

5 Explain how the braking distance, braking force and kinetic energy of the car in diagram D are related to 'work done'.

6 Produce a concept map about work.

Kinetic energy transfers

By the end of this topic you should be able to:

- describe what kinetic energy is
- explain examples of situations where kinetic energy is being transformed
- describe what elastic potential energy is
- **H** calculate kinetic energy.

Kinetic energy is the scientific name for movement energy. It can be transferred into other forms of energy, such as heat and sound. This happens when a car brakes. The kinetic energy is changed into heat and sometimes sound in the brakes and tyres.

1 What is kinetic energy?

2 a Give three examples of situations when kinetic energy is changed into another form of energy.

  b What forms of energy is it being transformed into in each situation?

A This car's kinetic energy was changed suddenly into other forms of energy.

B The kinetic energy of the wind is transformed into the kinetic energy of the boats.

Kinetic energy depends on the mass and speed of an object. For example, boat A will have more kinetic energy than boat B travelling at the same speed as boat A has more mass. If they both hit something, boat A will do more damage as it has more energy to transfer. If two identical boats sail at different speeds, the faster one will have more kinetic energy.

3 Chloe's car has more kinetic energy than Janice's. Describe two things that might be different between the two cars.

Elastic potential energy is stored in an object by doing work to change its shape. For example, when you bounce on a trampoline and it stretches, it stores elastic potential energy. You transfer energy to the trampoline by doing work on it. The elastic potential energy is released by transforming it into other forms. For example, it is transformed into kinetic energy when you bounce back up from the trampoline.

C

4 a Explain how the archer in photograph C stores elastic potential energy in the bow.
  b What energy transfer happens when the archer fires the arrow?

H

We can calculate the kinetic energy an object has using the equation:

kinetic energy = ½ ×   mass   ×   (speed)^2
    E_k = ½ ×   m   × v^2
  [joule, J]   =   [kilogram, kg)] × [(metre/second)^2, (m/s)^2]

Example

What is the kinetic energy of a 65 kg girl running at 6 m/s?

E_k = ½ × m × v^2
   = ½ × 65 × 6^2
   = ½ × 65 × 36
   = 1170 J

5 Calculate the kinetic energy of the following:
  a a 2 kg toy robot dog walking at 2 m/s
  b a boy on a bike riding at 8 m/s (The mass of the boy and his bike is 70 kg.)
  c a boat with a mass of 1200 kg moving at 5 m/s
  d a whale swimming at 7 m/s that overtakes the boat (The whale's mass is 4000 kg.)
  e the 25 kg of water that the whale squirts into the air at 11.5 m/s.

6 Brakes that overheat can fail. Write a short leaflet for a company that makes brake parts, to tell the public why brakes get hot and so what can be done to reduce this. Include explanations of:
  a how the brakes do work (including the equation for work)
  b how the stopping distance and force of the brakes are greater for greater speeds.

H Include in your leaflet the fact that F × d is equal to ½mv^2.

P

A

P

D

P

P

Momentum

By the end of this topic you should be able to:

- calculate momentum
- define momentum
- describe how forces affect the momentum of a body that is moving or able to move
- use an equation to show how force, change in momentum and time taken for the change are related.

An object which is moving has kinetic energy. It also has **momentum (plural: momenta)**. We can calculate the momentum using the equation:

$$\text{momentum} = \text{mass} \times \text{velocity}$$
$$p = m \times v$$

[kilogram metre/second, kgm/s] = [kilogram, kg] × [metre/second, m/s]

Example

What is the momentum of an 800 kg car moving at 12 m/s?

$p = m \times v$
$\quad = 800 \times 12$
$\quad = 9600$ kgm/s

1 Calculate the momentum of:
   a a 742 kg elephant moving at 5 m/s
   b a 70 kg woman skydiving at a terminal velocity of 53 m/s.

2 Would you prefer to try and stop the elephant or the skydiver? Explain your answer.

Momentum is a measure of how much something is moving. It has **magnitude** and direction. This means you must say how much momentum something has, and you must also say what direction the momentum is in.

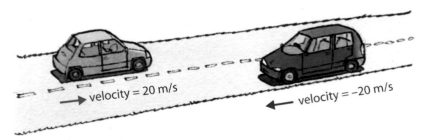

velocity = 20 m/s

velocity = −20 m/s

A Why have these cars got different momenta?

3 What part of the equation to calculate momentum tells us that we must think about its direction as well as the amount of momentum?

When a force acts on something that is moving, or able to move, there is a change in its momentum. The force needs to be larger to change momentum more quickly.

Newton originally worked out that force, change in momentum and time taken for the change in momentum are related:

$$\text{force} = \frac{\text{change in momentum}}{\text{time taken for the change}}$$

$$F = \frac{p_2 - p_1}{t_2 - t_1}$$

$$[\text{newton, N}] = \frac{[\text{kilogram metre/second, kgm/s}]}{[\text{second, s}]}$$

Example

What is the force needed to change a cyclist's momentum by 25 kgm/s in 5 seconds?

$$F = \frac{p_2 - p_1}{t_2 - t_1}$$

$$= \frac{25}{5}$$

$$= 5 \text{ N}$$

B Applying a force will change momentum.

4 Calculate the following:
  a the force needed to change a car's momentum by 3000 kgm/s in 5 seconds
  b the force needed to change a boat's momentum by 400 kgm/s in 8 seconds

5 A 75 kg man starts a sprint race. He increases his velocity from rest to 3 m/s in 1 second. What force must he be using to do this?

Car safety

C Why do you fly forwards if your car stops suddenly?

Seatbelts help to save people's lives in an accident. If you are not strapped in, your momentum keeps you moving if the car stops suddenly. You will continue to move forwards inside the car and could hurt yourself by hitting the dashboard or windscreen. The seatbelt keeps you in position. Losing your momentum very quickly needs a large force and this can injure you. In a sudden stop, the seatbelt stretches a little, reducing your momentum over a longer time, so the force on you is lower.

An airbag in a car can reduce your momentum to zero, so you stop moving. By doing this slowly, the force on you is less than if you hit the hard dashboard and stopped suddenly.

6 Write a short story about the Animalympics. Include details of the masses and velocities of five different animals in a race. Include at least one of the animals changing velocity and explain how this affects the momentum and how it might have happened. At the end, explain how to calculate the momenta of each animal and then put them in order of increasing momentum.

H Add a calculation of the force needed to cause the momentum change in your story.

Conservation of momentum

By the end of this topic you should be able to:

- describe what happens to momentum during a collision or an explosion when no external forces act
- use the conservation of momentum to calculate the mass, velocity or momentum of an object in a collision or explosion.

Moving objects have momentum. They may be involved in **collisions** with each other, or in explosions. In all cases, the total momentum in any situation is **conserved** as long as there are no other forces from outside acting on the objects involved.

This means that if you add up the total momenta before a collision or explosion, and the total momenta after the collision or explosion, they will be the same. The calculation must be done remembering to include the direction of the momenta.

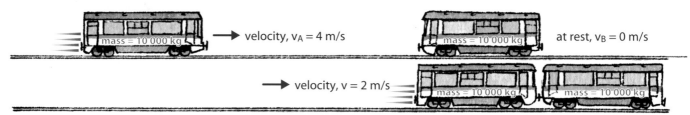

mass = 10 000 kg   velocity, v_A = 4 m/s   mass = 10 000 kg   at rest, v_B = 0 m/s

velocity, v = 2 m/s   mass = 10 000 kg   mass = 10 000 kg

A Momentum is conserved in any collision.

Example

Look at diagram A.

Momentum before collision:

$$p = (m \times v_A) + (m \times v_B)$$
$$= (10\,000 \times 4) + (10000 \times 0)$$
$$= 40000 + 0$$
$$= 40\,000 \text{ kgm/s}$$

Momentum after collision:

$$p = 2m \times v_{AB}$$
$$= 20\,000 \times 2$$
$$= 40\,000 \text{ kgm/s}$$

So, the momentum before the collision equals the momentum after the collision.

1 Explain what it means to say that 'momentum is conserved in a collision'.

2 Look at diagram B.
  a Calculate the momentum of each penguin before it collides.
  b Calculate the total momentum before the penguins collide.
  c Which direction is the total momentum in before the collision?
  d Calculate the total momentum after the collision and say in which direction it is.
  e Has the momentum been conserved?

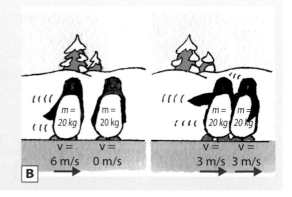

m = 20 kg   m = 20 kg   m = 20 kg   m = 20 kg

v = 6 m/s   v = 0 m/s   v = 3 m/s   v = 3 m/s

B

The forces that things exert on each other are equal and in opposite directions. For example, when the force of gravity of the Earth pulls a falling stone downwards, you can see it accelerate towards the Earth. The stone's gravity exerts an equal and opposite force on the Earth, but because the Earth is so massive, we cannot see its tiny acceleration caused by the stone's gravity force.

D Momentum is conserved in explosions.

An explosion might not be fast and violent. It could just mean two things that were together and not moving exert a force on each other to push apart. The momentum of each will be the same amount but in opposite directions. So, if you add up the momenta, accounting for the directions, the total will be zero, as it was before the explosion.

This idea is used to launch space rockets. The gases from the rocket engine are pushed downwards, and this makes an equal and opposite force which pushes the rocket upwards.

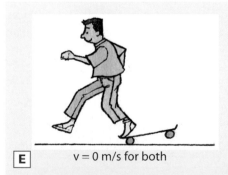

E   v = 0 m/s for both

4 Look at diagram E of Harry stepping off his skateboard.
   a Why will the skateboard move away from Harry?
   b What is the total momentum before Harry steps off?
   c Harry's mass is 50 kg. After stepping off, he is moving to the left at 2 m/s. What is his momentum?
   d What will be the momentum of the skateboard after Harry has stepped off it?
   e If the skateboard has a mass of 4 kg, how fast will it move?

weight = 700 N

force from horse = 700 N

C Forces always act in equal and opposite pairs.

3 a Give three examples of forces acting.
  b For each of your examples in a, write down where the equal and opposite force must act.

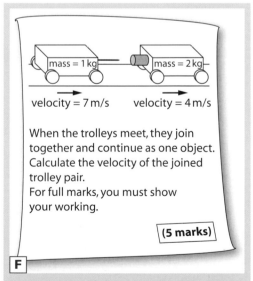

mass = 1 kg   mass = 2 kg

velocity = 7 m/s   velocity = 4 m/s

When the trolleys meet, they join together and continue as one object. Calculate the velocity of the joined trolley pair.
For full marks, you must show your working.

(5 marks)

F

5 Write a model answer to the exam question in F. It should tell examiners:
  a how many marks each part of the question is worth
  b what needs to be included in the answer to get the marks.

Atomic structure

P
H
P
P

By the end of this topic you should be able to:

- give relative masses and relative electric charges of protons, neutrons and electrons
- explain why atoms have no net electrical charge
- define mass number and atomic number and explain how this relates to elements
- explain how Rutherford and Marsden's work changed the accepted model of an atom.

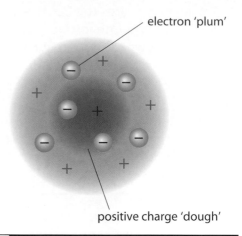

electron 'plum'

positive charge 'dough'

A The 'plum pudding' model of an atom.

Atomic models

Until the early twentieth century, **atoms** were described using the '**plum pudding model**'. **Electrons** (the 'plums') were spread randomly through a positive 'dough'. It was the best theory based on the evidence at the time.

1 Explain what is meant by the phrase 'plum pudding model of an atom'.

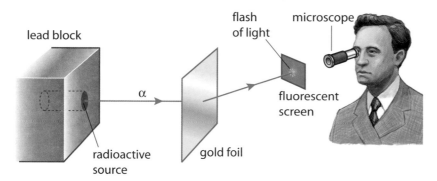

B Rutherford and Marsden's alpha scattering experiment.

In 1909, Lord Rutherford and two of his students, Geiger and Marsden, studied what happens when alpha particles hit a thin sheet of gold. They set up equipment with a radioactive source that fired **alpha (α) particles** at the gold. They measured how many went straight through the metal, and how many were deflected to different angles as they passed through the gold.

2 What was used as a target for the alpha particles in Rutherford and Marsden's alpha scattering experiment?

3 Alpha particles are absorbed by a few centimetres of air. How do you think Rutherford stopped the alpha particles being absorbed by air in the equipment?

	Observation	Conclusion about atoms
(i)	Most alpha particles went straight through the gold with no change in direction.	Most of the atom is empty space.
(ii)	Some alpha particles changed direction on their way through the atom.	The nucleus has a positive charge.
(iii)	A few alpha particles were repelled back the way they had come. (None were expected to do this.)	The small nucleus is central, has a large mass and a large positive charge.

D Conclusions from the alpha particle scattering experiments.

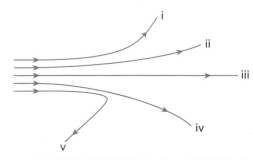

C Which way do alpha particles travel after passing through atoms of gold?

4 a What did Geiger and Marsden see happening in Rutherford's experiment that meant that most of an atom must be empty space?

b Why would a positive alpha particle be repelled from a nucleus?

c Why must a nucleus have a large mass to repel an alpha particle straight back? (*Hint*: think about momentum.)

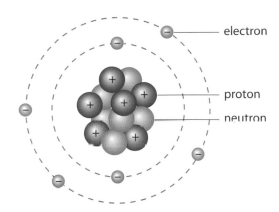

electron

proton

neutron

E Rutherford's idea of a carbon atom.

Now we think of an atom as having a **nucleus** made of **protons** and **neutrons**. The number of protons in the nucleus tells us what **element** the atom is, because all atoms of a particular element have the same number of protons. Different elements have different numbers of protons. The total number of protons is called the **atomic number**. The number of electrons is equal to the number of protons, giving the atom no net electrical charge. The total number of neutrons and protons in an atom's nucleus is called its **mass number**.

5 Lithium has three protons in its nucleus, and a mass number of seven. Draw a diagram showing the structure of an atom of lithium.

Particle	Mass	Charge
Electron	almost zero	−1
Neutron	1	0
Proton	1	+1

F Charge and mass properties of the parts of atoms.

6 Write a script for a scene in a play in which Lord Rutherford meets J. J. Thomson. Thomson has heard that Rutherford wants to replace the plum pudding model. Your script should include Rutherford explaining:

a how his experiment was set up

b what observations were made

c how these observations led to the conclusion that atoms must have a small, positive nucleus with all the mass.

Radioactivity

By the end of this topic you should be able to:

- explain how ions are produced
- define what isotopes are
- explain where background radiation comes from
- describe how radioactivity affects nuclei.

Some nuclei have different numbers of neutrons but the same number of protons. They are nuclei of the same element because they have the same number of protons. Such nuclei are called **isotopes**.

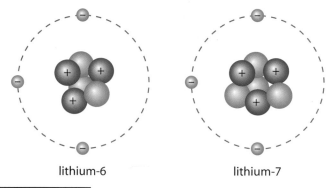

lithium-6   lithium-7

A Lithium isotopes.

Most elements have stable isotopes. Some have unstable isotopes where the number of neutrons makes the nucleus unstable. If an unstable radioactive nucleus gives off an alpha particle, two protons and two neutrons leave the nucleus. The alpha particle is two protons and two neutrons moving together. The isotope's proton number goes down by two and its mass number goes down by four. As the nucleus now has fewer protons, it has become a different element.

americium-241 nucleus

alpha particle

B A nucleus losing an alpha particle.

1 a How are the isotopes lithium-6 and lithium-7 different?
 b How are they the same?

2 Look at diagram B. If a nucleus of an atom gives off an alpha particle, what will happen to:
 a the proton number
 b the neutron number
 c the electron number?

3 Why does giving off an alpha particle make an element change into another element?

Beta decay makes a neutron in the nucleus change into a proton and an electron. The electron flies off and we call this a **beta (β) particle**. The proton stays in the nucleus. Since the nucleus now has one more proton, it is a different element.

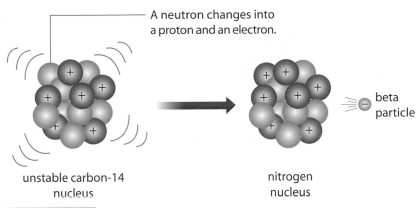

A neutron changes into a proton and an electron.

beta particle

unstable carbon-14 nucleus

nitrogen nucleus

C Beta decay.

Atoms may lose or gain electrons to form charged particles called **ions**.

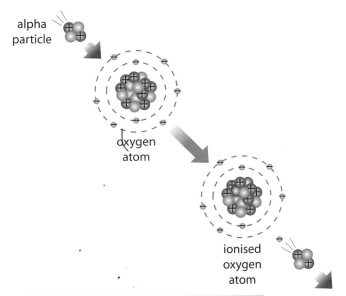

alpha particle

oxygen atom

ionised oxygen atom

D Radioactive particles can ionise other atoms.

When alpha or beta particles pass through air, or any material, they **ionise** the atoms of that material. Electrons from atoms of the material are forced away from their atoms by near collisions with the charged radioactive particles.

6 If a beta particle ionises an oxygen atom, what will happen to the following:
a the oxygen's proton number
b the oxygen's neutron number
c the oxygen's electron number?

7 Why does being ionised not make an element change into another element?

4 If a nucleus of an atom gives off a beta particle, what will happen to:
a the proton number
b the neutron number
c the electron number?

5 Why does giving off a beta particle make an element change into another element?

8 Plutonium-241 has 94 protons and 147 neutrons. A proton number of 92 gives the element uranium, and a proton number of 95 is americium.
a Draw a sketch of a plutonium-241 nucleus giving off:
  (i) an alpha particle
  (ii) a beta particle.
b Explain underneath your sketches what changes would happen in the plutonium nucleus and what elements would be produced.
c Show the alpha and beta particles ionising nitrogen atoms nearby.

Nuclear fission and fusion

By the end of this topic you should be able to:

- describe what nuclear fission is and how it happens
- give two examples of metals used in nuclear reactors
- describe what nuclear fusion is and where it happens naturally
- sketch diagrams to show how fission and fusion can start chain reactions.

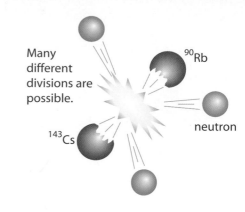

fission of uranium-235

A Fission is the splitting of an atomic nucleus.

Very large, unstable, nuclei can split into pieces. This is **nuclear fission**. It produces two smaller nuclei, two or three neutrons, and also releases energy. The smaller nuclei produced are more stable than the one at the start.

1 a In diagram A, a uranium-235 nucleus splits into rubidium-90, caesium-143 and three neutrons. What is this process called?
  b What else is also released in this reaction?

The most common fissionable substances used in nuclear reactors are **uranium-235** and **plutonium-239**. For fission to occur easily in the uranium-235 or plutonium-239 nucleus, it must first absorb a neutron.

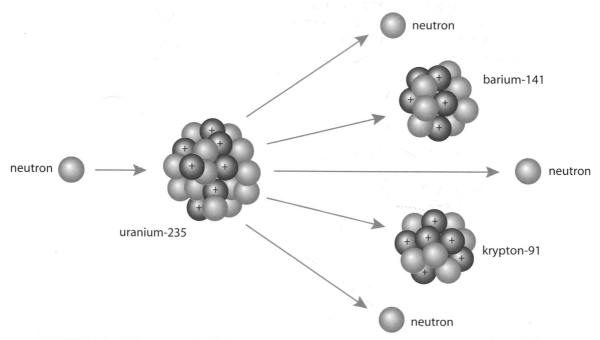

B Some nuclei must absorb an extra neutron before fission will occur.

Fission produces extra neutrons, so if we can get uranium-235 or plutonium-239 nuclei to split, the neutrons produced could split another nucleus. This fission would then produce more neutrons, making more nuclei split. This is called a **chain reaction**.

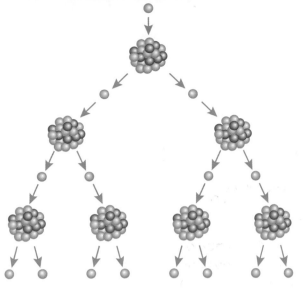

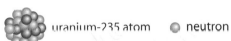
uranium-235 atom   ○ neutron

C How could a chain reaction run out of control, as in a nuclear explosion?

This chain reaction won't continue without our help. The neutrons released in fission are high energy, and so move very fast. Uranium-235 nuclei can't easily absorb the extra neutron needed for fission if the neutron is moving at high speed.

In a power station reactor, the neutrons are made to pass through a material which slows them down to a better speed. This **moderator** is usually made of graphite or water. At Sellafield, there was a nuclear reactor in which uranium-235 was used in a fission chain reaction until it was decommissioned in 2003. Plutonium-239 doesn't need a moderator.

4 In a nuclear power station, what is the job of the moderator?

5 Name two materials that can be used as a moderator.

Nuclear fusion also releases energy. In the fusion process, two small nuclei join together to make a larger one. For example, two hydrogen isotopes may join up to become a helium nucleus. It is not easy to get nuclei to combine and the temperature needs to be high to make the reaction happen.

Energy is released in stars by the process of nuclear fusion. Stars start off their lives as mostly hydrogen. As nuclear fusion reactions go on, the hydrogen is slowly used up. After a time, the helium nuclei produced may combine in nuclear fusion, producing other elements and more energy. Eventually, the star cannot join the elements further and the nuclear fusion stops. The star has run out of energy.

6 Where do stars get their energy from to shine so brightly?

2 Draw a sketch of how a chain reaction could happen with plutonium-239 nuclei.

3 Why would a chain reaction produce a huge amount of energy?

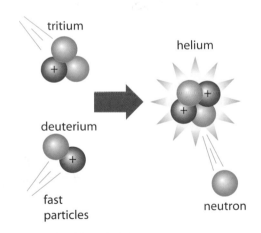

D Nuclear fusion gives off the energy that powers stars.

7 Fission and fusion are often confused by students. Make a revision card to explain the differences. Your card should have fission on one side, including chain reactions and nuclear power stations. On the other side you should include fusion and how it can produce energy in stars.

Investigative Skills Assessment 1

The Safachutes Company makes parachutes. The company need to know the terminal velocity of their parachutes in order to make sure they are safe. To check the terminal velocity, the company has tested their parachutes in a vertical wind tunnel.

Using four people with different masses, several different sized parachutes were tested in the wind tunnel. The terminal velocity was found at the wind speed that held the parachute exactly stationary in the tunnel, so the weight force equalled the air resistance.

Here are the results:

Area of parachute (m²)	Terminal velocity for different skydiver masses (m/s)			
	50 kg	60 kg	70 kg	80 kg
10.0	3.16	3.46	3.74	4.00
20.0	2.24	2.45	2.65	2.83
30.0	1.83	2.00	2.16	2.31
40.0	1.58	1.73	2.87	2.00
50.0	1.41	1.55	1.67	1.79
60.0	1.29	1.41	1.53	1.63
70.0	1.20	1.31	1.41	1.51
80.0	1.12	1.22	1.32	1.41
90.0	1.05	1.15	1.25	1.33
100.0	1.00	1.10	1.18	1.26

Study the results table carefully and then answer these questions.

1 a What was the range of areas used? *(1 mark)*
  b From the results in the table, what was the precision of the area measurement? *(1 mark)*

2 Which one of the following was a control variable in this experiment? Choose from: the mass of the person; the wind speed; the area of the parachute; none of these. *(1 mark)*

3 a There is one anomalous result in the tables. State the value of this anomalous result. *(1 mark)*
  b How should this anomalous result be dealt with? *(1 mark)*

4 Why do you think that they did not test for an area of 0 m²? *(1 mark)*

5 Why did the company test for different mass skydivers instead of just one mass in these experiments? Choose from: to improve accuracy; to improve sensitivity; to improve precision; to improve validity. *(1 mark)*

6 a Compare the results for different mass skydivers. What can you say about the terminal velocities for a skydiver with greater mass? *(1 mark)*
  b What do the results tell you about how terminal velocity changes with the area of a parachute? *(1 mark)*

7 a The maximum safe landing velocity for a person on a parachute is 3 m/s. Write a warning statement for people wanting to buy a parachute which tells them what size of parachute to avoid so they won't land too quickly. *(2 marks)*
  b Explain how the experiments support your safety statement. *(1 mark)*
  c Add a safety statement about the mass limit of a person who could safely use a Safachute. *(1 mark)*
  d ✎ Explain how the experiments support this safety statement. *(3 marks)*

8 How could the Safachutes company improve the experiment to make it useful for a wider range of customers? *(1 mark)*

Total = 17 marks

Electricity and circuits

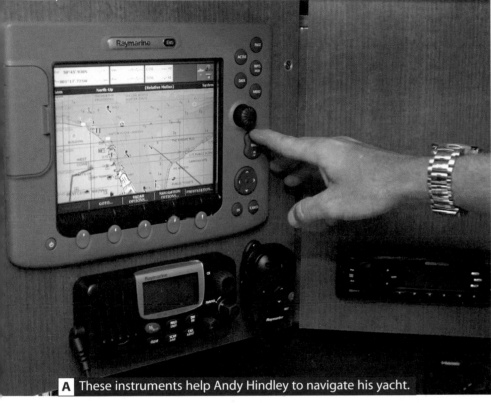

A These instruments help Andy Hindley to navigate his yacht.

Electricity is a very useful way of transferring energy. Some energy transfers can also be done without using electricity. Instead of using an electric fire to heat a room, you could burn a fuel. There are some things that only be done using electricity.

A yacht uses electricity from a battery when it is at sea. The battery is recharged when the engine runs. When the yacht is in harbour it needs to be connected to the mains supply.

By the end of this unit you should be able to:

H

- describe the dangers and uses of static electricity
- recall and use equations relating charge, current, energy, potential difference and time
- describe how the current through a resistor, diode and filament lamp changes when the potential difference changes
- explain how the resistance of LDRs and thermistors changes, and suggest how these can be used in circuits
- describe the differences between direct and alternating current
- describe some of the safety features used in domestic wiring and appliances.

1 a Write a list of five machines that transfer electrical energy into other forms of energy.
  b For each machine in your list, say whether another form of energy could be used instead of electricity.

2 List three items that:
  a use electricity from batteries
  b use electricity from the mains supply
  c can use electricity from either source.

3 The wiring in buildings and appliances is safe if it is used sensibly. What three features are included in wiring and appliances for safety?

Static electricity

By the end of this topic you should be able to:

- explain how rubbing insulating materials gives them an electrical charge
- explain positive and negative charges in terms of the movement of electrons
- describe the forces between bodies with the same type of charge and between bodies with different types of charges.

If you have ever had a shock when getting out of a car, or heard your clothing crackle as you pull it over your head, you have experienced an effect of **static electricity**.

You can give many **insulating materials** a **charge** of static electricity by rubbing them. This is sometimes called an **electrostatic charge**.

A This girl is charged with static electricity.

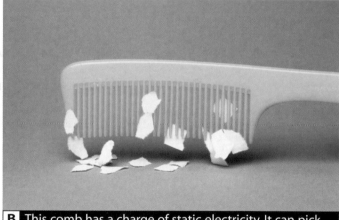

B This comb has a charge of static electricity. It can pick up small pieces of paper.

1 Describe two things that static electricity can do.

2 What kind of materials can be given a charge?

All atoms contain electrically charged particles called **protons** and **electrons**. Protons have a positive charge, and are found in the **nucleus** of atoms. Electrons move around the nucleus of the atom. An atom normally has no overall charge because the positive charges on the protons are balanced by the negative charges on the electrons.

3 Where are these particles found in an atom:
   a proton   **b** electron?

4 An atom has 10 protons.
   a How many electrons does it normally have?
   b Explain your answer.

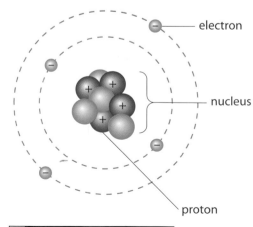

C The structure of an atom.

If you rub two insulating materials together, electrons may be transferred from one material to the other. Protons cannot be transferred, because they are fixed in the nuclei of atoms.

When you rub an acetate rod with a piece of cloth, some of the electrons in the acetate move onto the cloth.

The acetate now has more protons than electrons, so it has a positive charge. The cloth has more electrons than protons, so it has a negative charge.

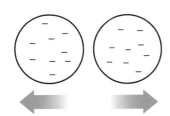
When you rub a polythene rod, some of the electrons in the cloth move onto the polythene.

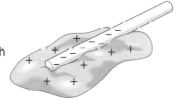

D The effect of rubbing two insulating materials together.

5 a Which particles can be transferred when you rub an insulating material?
b Explain your answer.

6 Explain why a polythene rod gets a negative charge when you rub it.

When you bring two objects with static charges close to each other, they will exert a force on each other.

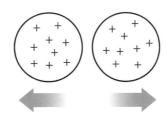
If the charges are the same as each other, the two objects **repel** each other (push each other apart).

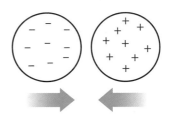
If the charges on the objects are different, the two objects **attract** each other.

E Attraction and repulsion.

7 Look at photograph A. Explain why the girl's hair is standing up.

8 You rub two balloons on your jumper. Explain why the balloons will then repel each other. Include the following words in your answer: proton, electron, charge, repel.

The dangers of static electricity

By the end of this topic you should be able to:

- explain how a charged body can be discharged
- explain how static electricity can cause sparks
- describe some situations in which a static charge can be dangerous, and how precautions can be taken to make sure it is discharged safely.

Static electricity can build up on clouds, and can cause a spark to form between the cloud and the Earth. We see these sparks as lightning. The lightning also makes sound waves that we hear as thunder.

Lightning is just one example of the possible dangers of static electricity. Buildings can be protected from lightning using a lightning conductor. This is a strip of conducting material down the side of the building that allows the lightning to travel safely through it instead of through the building.

If you are in a car during a thunderstorm you will be safe. If lightning hits the car, it is easier for the electricity to travel through the metal body of the car than to go through you, and so you will not be harmed.

H Lightning happens because there is a large **potential difference** between the cloud and the Earth. The potential difference is a measure of the energy that could be released when the charge moves. The greater the charge, the higher the potential difference.

Many people think that a car will protect you in a thunderstorm because the electricity that causes the lightning cannot flow through the rubber tyres. However, the potential difference in a lightning strike is high enough to allow electricity to flow through the air between the car body and the ground, and it can also flow through any water on the outside of the tyres.

A Lightning can damage buildings and trees, and can kill people and animals.

1 **a** What causes lightning?
  b Why is lightning dangerous?

2 **a** How can buildings be protected against lightning?
  b Explain how this protection works.

3 Why is a car a safe place to be in a thunderstorm?

You can sometimes build up an electrostatic charge by walking on a carpet or just from sitting on a chair. You only find out that you have become charged when you touch a metal object such as a doorknob. You feel a small electric shock when the charge on you flows through the object.

B Charge builds up as your shoes rub on the carpet. The carpet is an insulating material, so the charge cannot flow anywhere.

C Charge can flow from you, through the metal door to earth.

D You are **discharged**.

Static electricity is only dangerous when it is likely to cause a spark. If a conducting path can be provided to discharge the static electricity, then there will be no spark. People making or mending computers use an antistatic wrist strap connected to the casing. This stops static discharges damaging the components which may be destroyed by a spark.

bonding line

E Static electricity can build up as fuel flows through a refuelling pipe, and aircraft can also build up a static charge as they move through the air. A bonding line (a metal wire) is used to connect the aircraft to earth before it is refuelled.

4 Why do people working on electronic equipment wear wrist straps that connect them to the equipment?

5 a How could a spark cause an accident when an aircraft is being refuelled?
b How does the bonding line prevent this?
c Why is it important that the bonding line is connected before the fuel nozzle touches the aircraft?

6 Describe some of the dangers of static electricity and how electrostatic charges can be discharged safely. Present your answer as a series of cartoons, a poster or a short paragraph.

Using static electricity

By the end of this topic you should be able to:

- explain ways in which static electricity can be useful
- describe the basic operation of photocopiers and smoke precipitators.

Static electricity can be dangerous because sparks can cause fires or explosions. It can also be very useful. It can help to make painting things more efficient.

B Electrostatic charges can be used to reduce the amount of pollution released into the atmosphere.

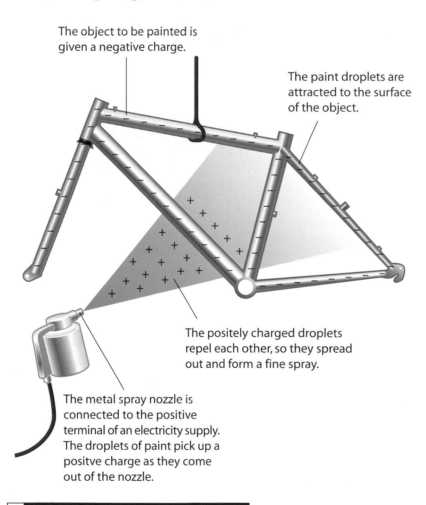

The object to be painted is given a negative charge.

The paint droplets are attracted to the surface of the object.

The positely charged droplets repel each other, so they spread out and form a fine spray.

The metal spray nozzle is connected to the positive terminal of an electricity supply. The droplets of paint pick up a positve charge as they come out of the nozzle.

A How electrostatic paint spraying works.

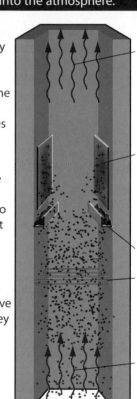

The negatively charged dust particles are attracted to the positively charged plates and stick to them.

The plates are tapped occasionally to make the dust fall into hoppers.

Dust particles collect negative charges as they flow past the wires.

smoke with dust particles removed

positively charged collector plate

hopper

wires with a negative charge

smoke with dust particles

C How a smoke precipitator works.

1 Why do the drops of paint spread out when they leave the nozzle?

2 Why does the object have to be given a negative charge?

Smoke precipitators use static electricity to remove dust particles from the smoke coming out of a chimney.

3 What is an electrostatic precipitator used for?

4 Look at diagram C.
  a How do the dust particles get charged?
  b Why do the collector plates need to have a positive charge?

Static electricity is also used in **photocopiers**. Diagram D describes how photocopiers work.

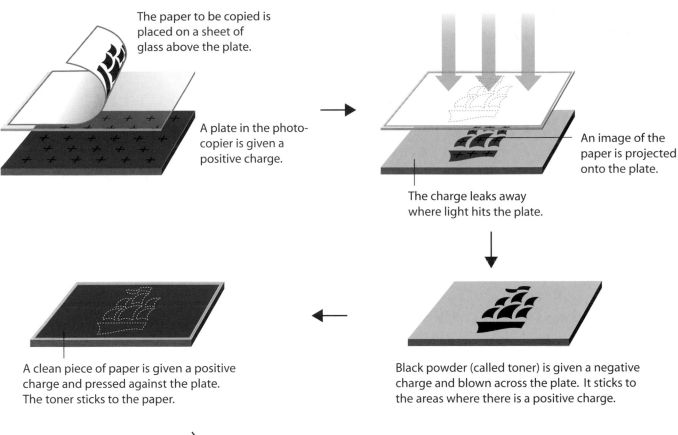

The paper to be copied is placed on a sheet of glass above the plate.

A plate in the photo-copier is given a positive charge.

An image of the paper is projected onto the plate.

The charge leaks away where light hits the plate.

A clean piece of paper is given a positive charge and pressed against the plate. The toner sticks to the paper.

Black powder (called toner) is given a negative charge and blown across the plate. It sticks to the areas where there is a positive charge.

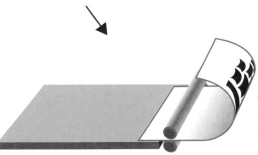

The paper with the toner passes through rollers which heat it to make the toner stick to the paper permanently.

D How a photocopier works.

5 a How does a static charge help to make sure that the toner powder spreads out onto the plate?
 b Why does the toner powder only stick to parts of the plate?

6 Why does the clean piece of paper need to be given a positive charge?

7 Why does a piece of photocopied paper feel warm when you take it out of the machine?

8 Write two sets of bullet points to explain how static electricity is used in:
 a smoke precipitators
 b photocopiers.

Electric current

By the end of this topic you should be able to:

- explain that electrical charges can move through some materials such as metals, and that this movement is called a current
- use symbols to show components in circuits
- **H** use the equation: charge = current × time.

Electrical charges can move easily through some materials. These materials are called **conductors**. Metals are good conductors of electricity because some of the electrons in them can move between atoms. This movement causes a flow of charge called a **current**. In plastics and other insulating materials the electrons cannot move around easily.

Free electrons move around all the time in different directions in metals, but they are not actually going anywhere. When we put a metal wire in a circuit, the **cell** creates a force on the free electrons in one direction. This causes a current to flow around the **electric circuit**.

1 What is the difference between conductors and insulators? Use the word 'electrons' in your answer.

2 Give two examples of electrical conductors.

3 What is needed to push electrons around a circuit?

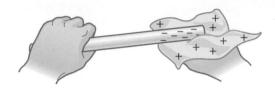

Electrons are transferred from the cloth to the polythene rod. They cannot move through the polythene, so the end of the rod has a static charge.

Electrons are transferred from the cloth to the metal rod. Metals conduct electricity, so the extra electrons spread themselves out through the metal. It is difficult to detect the extra static charge.

A Why it it difficult to charge a metal rod with static electricity.

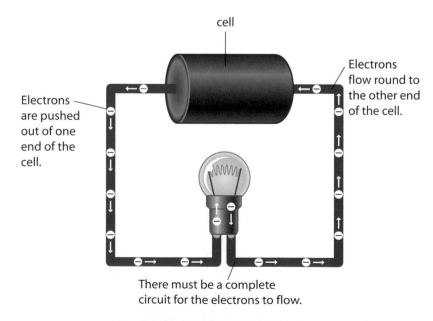

cell

Electrons are pushed out of one end of the cell.

Electrons flow round to the other end of the cell.

There must be a complete circuit for the electrons to flow.

B In an electric circuit there is a continuous path made from conducting materials.

You cannot see charges moving when a current flows in a circuit so we need a measuring instrument. An **ammeter** is used to measure current and the unit of current is the **ampere** (**A**) (or 'amp'). Very small currents are measured in **milliamps** (**mA**). 1 mA = 1/1000 A.

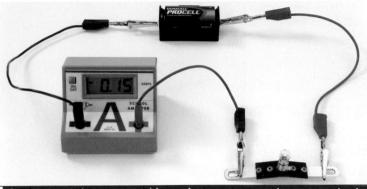

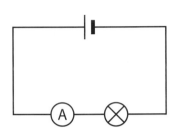

C An ammeter is connected **in series** to measure the current in the circuit.

The charge is not 'used up' as it moves around the circuit, and so the current in a simple circuit is the same everywhere.

4 Why do we need to use an ammeter to measure the current in a circuit?

5 Ammeter A1 in circuit D reads 0.5 A. What is the reading on:
 a ammeter A2
 b ammeter A3?

6 Explain how you worked out your answers to question **5**.

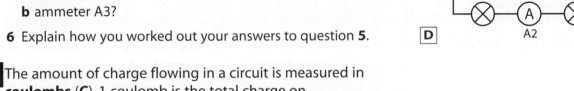

D

H The amount of charge flowing in a circuit is measured in **coulombs** (**C**). 1 coulomb is the total charge on 6 241 000 000 000 000 000 electrons (you do not need to remember this number!).

The charge and the current are linked by the following equation:

charge   =   current   ×   time
(coulomb, C)   (ampere, A)   (second, s)

Example

A current of 5 A flows for 10 seconds. How much charge has flowed through the circuit?

charge = current × time
charge = 5 A × 10 s
     = 50 C

6 Look at circuit D again. The circuit is left switched on for 50 seconds.
 a How much charge flows through ammeter A1 in that time?
 b How much charge flows though ammeter A2?

7 Write a sentence for each of these words to explain what they mean:
 a charge
 b conductor
 c current
 d circuit.

H **8** Explain how you can calculate the amount of charge flowing in a circuit.

Transforming energy

By the end of this topic you should be able to:

P
- explain that electrical energy is transformed into heat energy when a current flows through a resistor
- explain what potential difference means in terms of electrons flowing in an electric circuit

H
- use the equation: energy transformed = potential difference × charge.

An electric current is a flow of charge. When a current flows through **components** like lamps, energy is transferred to the component. In resistors, the energy is transferred to heat energy. In lamps, the wire inside the lamp gets so hot that it glows, and some of the energy is transferred to light energy.

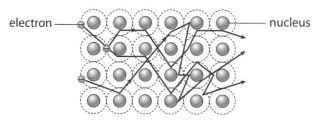

electron — nucleus

A Current flowing through a wire.

You can find out how much energy is being transferred in a component using a **voltmeter**. The voltmeter compares the energy carried by the current going into a component with the energy carried by the current leaving it. The change in energy is due to electrical energy being transferred to other forms in the components. This difference in energy is called the **potential difference**. The units for potential difference are **volts (V)**.

You can also use a voltmeter to work out how much energy a cell or battery gives the electrons as it pushes them around the circuit.

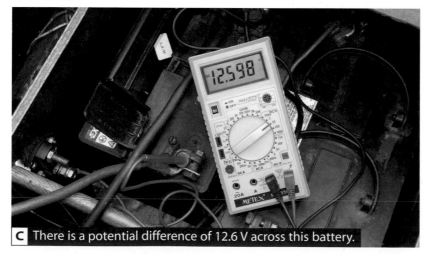

C There is a potential difference of 12.6 V across this battery.

1 Why do components get hot when electric charges flow through them?

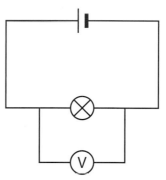

B The voltmeter is connected **in parallel** to the lamp to measure the potential difference across it.

2 a What does potential difference mean?
b What are the units for measuring potential difference?

You can use models to help you to think about what happens in circuits.

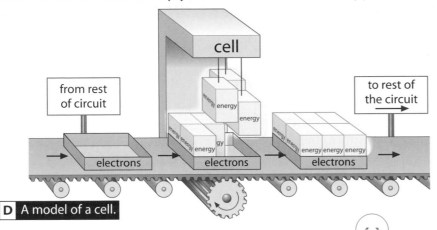

D A model of a cell.

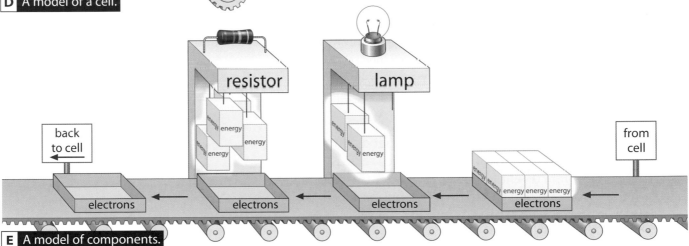

E A model of components.

▶P

3 Look at diagrams D and E. Which part of the model represents:
a a cell
b the wires
c the electrons?

4 How does the model show that charge is not used up as it moves around a circuit?

H The energy transferred in a component depends on:
- the potential difference (how much energy has been transferred by the charge)
- the charge that has flowed through the component.

---

energy transformed = potential difference × charge
 (joules, J)   (volts, V)   (coulombs, C)

Example

10 C of charge flows through a resistor. There is a potential difference of 4 V across the resistor. How much energy is transferred?

energy transferred = potential difference × charge
energy transferred = 4 V × 10 C
                          = 40 J

---

5 A light in the yacht has a potential difference of 12 V across it. How much energy is transferred when 1000 C of charge flows through it?

6 6000 C of charge flows through a yacht battery. The potential difference of the battery is 24 V. How much energy is transferred to the charge?

H

7 a What does the potential difference across a battery represent?
b What does the potential difference across a component represent?
c How do you calculate the energy transferred from potential difference and charge?

▶P ▶D ▶P ▶P

Resistance

By the end of this topic you should be able to:

- recall that the current through a resistor is proportional to the potential difference across the resistor
- use the equation: potential difference = current × resistance
- find the resistance of a component by measuring the current and potential difference
- explain how the current through a component depends on its resistance.

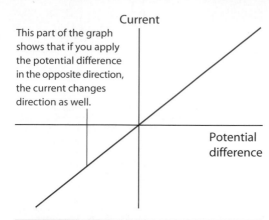

This part of the graph shows that if you apply the potential difference in the opposite direction, the current changes direction as well.

A The graph of current against potential difference is a straight line for a resistor at constant temperature.

The size of the current that flows through a resistor depends on the potential difference across the resistor. The bigger the potential difference, the higher the current. If the resistor is kept at a constant temperature, then the current is proportional to the potential difference. If you double the potential difference, you double the current.

The current through a component also depends on the **resistance** of the component. The resistance tells us how easy or difficult it is for current to flow. The greater the resistance, the harder it is for electricity to flow through the component and the smaller the current.

1 The potential difference in circuit B is 3 V. What will the current be if the potential difference is changed to:
 a 6 V
 b 12 V?

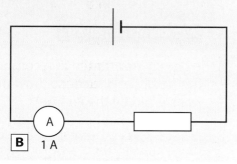

B 1 A

2 a Put the resistors R1, R2 and R3 in order of their resistance, smallest to largest.
 b Explain how you worked out your answer.

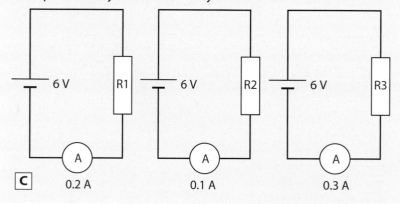

C 0.2 A   0.1 A   0.3 A

Graph D shows current–potential differences for three different resistors. Resistor X has the highest resistance. It has the smallest current for a particular potential difference.

3 a Explain how you can tell from graph D that resistor Z has the lowest resistance.
 b Copy and complete this sentence: The steeper the line on a current–potential difference graph, the _____ the resistance.

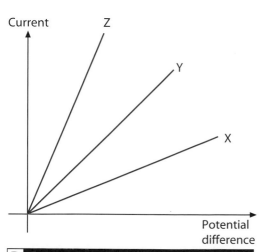

D Current–potential difference graphs for three different resistors.

You can find the resistance of a component by measuring the current through it and the potential difference across it. The units for resistance are **ohms** (Ω).

E Finding the resistance.

The resistance, potential difference and current are related by this equation:

potential difference = current × resistance
    (volt, V)   (ampere, A)   (ohm, Ω)

This equation can be rearranged to find the resistance if you know the current through a component and the potential difference across it:

$$\text{resistance} = \frac{\text{potential difference}}{\text{current}}$$

Example

Calculate the resistance of resistor R1 in diagram C.

$$\text{resistance} = \frac{\text{potential difference}}{\text{current}}$$
$$= \frac{6\,V}{0.2\,A}$$
$$= 30\,\Omega$$

4 Calculate the resistances of resistors R2 and R3 in diagram C.

5 **a** A resistor is connected to a 10 V supply, and a current of 0.5 A flows. What is its resistance?
  b The potential difference is reduced to 5 V. What will happen to the current?

6 The following measurements were taken for different resistors. Calculate the resistance of each one:
  a 12 V, 2 A
  b 230 V, 10 A
  c 2.5 A, 10 V.

7 **a** How does the resistance of a component affect the current flowing through it?
  b How can you measure the resistance of a component?

Current in circuits

By the end of this topic you should be able to:

- recall that the same current flows through each component in a series circuit
- work out the total resistance of a circuit with resistors connected in series
- recall that in a parallel circuit the total current is the sum of the currents through the separate branches.

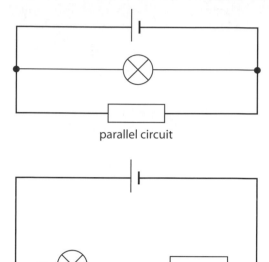

parallel circuit

series circuit

A Series and parallel circuits.

Lots of components can be connected together in a circuit. When all the components are connected in one 'loop', the circuit is a **series circuit**. If there is more than one 'branch' in a circuit, it is a **parallel circuit**.

In a series circuit there is just one path for the electric current. The same charged particles travel through the whole circuit, and so the current is the same everywhere in the circuit.

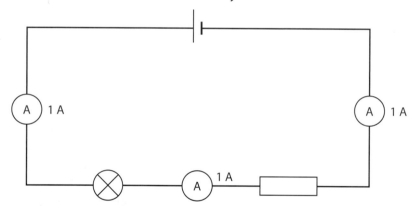

B In a series circuit the current is the same at any point in the circuit.

Rule 1

In a series circuit the current is the same at any point in the circuit.

1 The current leaving one end of a cell is 3 A. What will be the current returning to the other end of the cell?

The current in a series circuit depends on the overall resistance of the circuit and the potential difference provided by the cells. The higher the resistance of the circuit, the lower the current, as long as the potential difference is kept the same.

If there is more than one component, each component resists the flow of electricity. When components are added to a circuit in series, the resistance increases because the electrons have to be pushed through all of the components, one after the other.

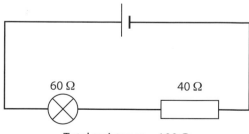

Total resistance = 100 Ω

C The total resistance in a series circuit is the sum of the resistances of all the components in the circuit.

Rule 2

The total resistance in a series circuit is the sum of the resistances of all the components in the circuit.

2 A resistor is added to a circuit in series. What happens to:
  a the resistance of the circuit
  b the current in the circuit?

3 a What is the total resistance in circuit D?
  b What current will flow through the 2 Ω resistor? Explain your answer.

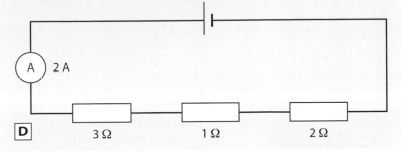

In a parallel circuit there is more than one path for the electric current to flow through. In circuit E the branches are not the same so different amounts of current flow down each branch, but the total current in all the branches is the same as the current flowing in and out of the cell.

Rule 3

In a parallel circuit the currents through each branch add up to the total current from (and back to) the cells.

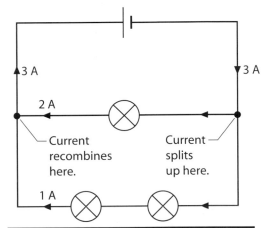

E In a parallel circuit the currents through each branch add up to the total current from the cells.

4 Look at circuit F.
  a What current is flowing from the cell?
  b How much current flows through the lamp?
  c Which meter measures the current returning to the cell?
  d What is the reading on this ammeter?

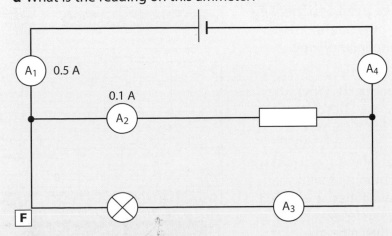

5 a Look at circuit D. How would removing the 3 Ω resistor change the overall resistance of the circuit?
  b Look at circuit F. Why is the current flowing through the cell different to the current flowing through the resistor? Explain your answer.

Potential difference in circuits

By the end of this topic you should be able to:

- recall that the potential difference of cells connected in series is the sum of the potential differences of each cell
- explain how the potential difference is shared between the components in a series circuit
- recall that the potential difference across each branch of a parallel circuit is the same.

Several cells can be used together in a circuit. A group of cells used together like this is called a **battery**. The potential difference provided to the circuit by a battery is the sum of the potential differences provided by all the cells, as long as they are connected in series, and all the same way round.

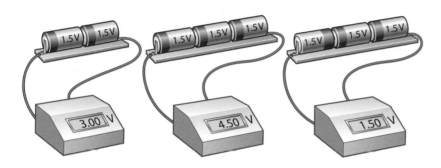

1 What potential difference is provided by these cells?

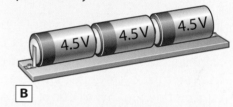

B

A Cells must all be connected the same way round to increase the potential difference in a circuit.

Rule 4

The potential difference provided by cells connected in series is the sum of the potential differences of all the cells, as long as they are all connected the same way round.

The electrical energy provided to a circuit by a cell or battery is converted to other forms of energy by the other components. The potential difference is a way of measuring the energy carried by the charge.

The cell in circuit C has a potential difference of 6 V. When the cell is in a circuit the energy it provides is shared between all the components. In circuit C the lamps are identical, so the potential difference across each one is 3 V. In circuit D the components are different and the potential difference across each one depends on its resistance. The component with the highest resistance has the highest potential difference across it. The potential differences across the components always add up to the potential difference supplied by the cell.

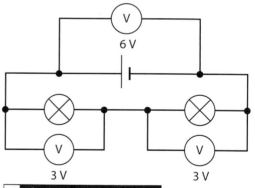

C The lamps are identical.

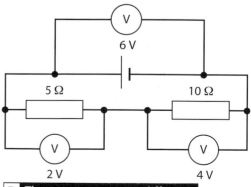

D The components are different.

Rule 5

In a series circuit the potential difference is shared between the components in the circuit.

2 What is the potential difference provided by the cell in circuit E?

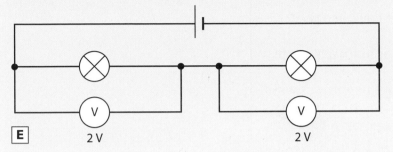

E   2 V   2 V

3 a What is the potential difference across resistor R2 in circuit F?
  b Which resistor has the highest resistance? Explain your answer.

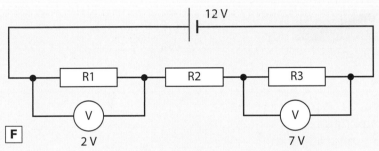

12 V

F   2 V   7 V

The charges leaving the cell transfer all their energy to the components in the circuit. It does not matter which route they take. This means that the potential difference across each branch of the circuit must be the same.

Rule 6

In a parallel circuit the potential difference is the same across each route.

4 a What are the voltage readings on voltmeters V1 and V2 in circuit H?
  b Explain your answer in terms of energy and electrons.

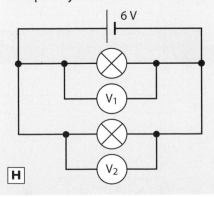

6 V

H

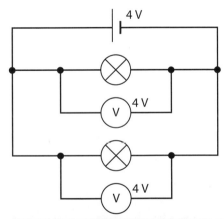

4 V

4 V

4 V

G In a parallel circuit the potential difference is the same across each route.

5 Look back at the circuit models in diagrams D and E on page 199.
  a Explain how this model illustrates Rule 5.
  b How would you change this model to help you to think about Rules 4 and 6?

Changing the resistance

By the end of this topic you should be able to:

- recall that the resistance of a filament lamp increases as the temperature of the filament increases
- recall that the current through a diode flows in one direction only
- sketch current–potential difference graphs for a filament lamp and a diode
- describe how the resistance of a thermistor changes with temperature
- recall that the resistance of an LDR decreases as light intensity increases.

A resistor is a component that controls the size of the current in a circuit. A high resistance makes it difficult for the current to flow, so there is only a small current. Engineers can work out the resistance needed in a circuit using the formula:

potential difference = current × resistance

This formula only works if the resistor stays at a constant temperature. If the component changes temperature, the resistance may change. Photograph A shows some apparatus used to investigate how the resistance changes in a **filament lamp** when the potential difference changes. Graph B shows the results of the investigation.

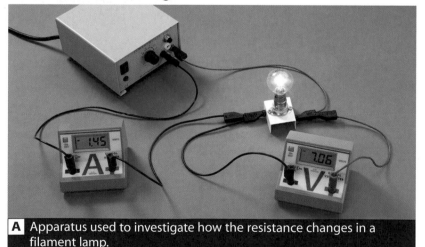

A Apparatus used to investigate how the resistance changes in a filament lamp.

As the potential difference and current increase, more energy is transferred and the filament gets hotter.

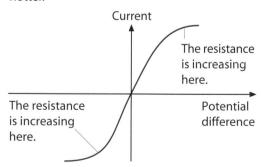

B Current–potential difference graph for a filament lamp.

1 **a** Is the temperature of the filament lamp highest when the potential difference is high or low?
   b Explain your answer.

2 **a** Does the resistance of the filament lamp increase or decrease when it gets hotter?
   b How can you tell this from graph B?

Some components are designed to change their resistance in certain conditions.

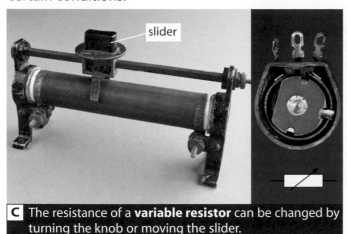

C The resistance of a **variable resistor** can be changed by turning the knob or moving the slider.

D A **thermistor** is a component whose resistance is designed to decrease when the temperature increases.

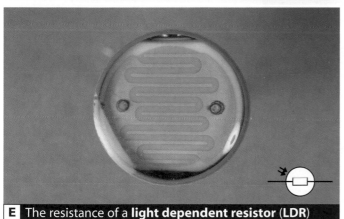

E The resistance of a **light dependent resistor** (**LDR**) decreases as light intensity increases (gets brighter).

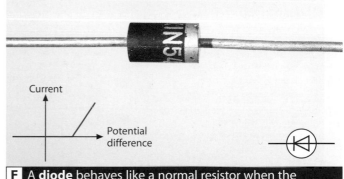

F A **diode** behaves like a normal resistor when the current flows in one direction. When a potential difference is applied in the other direction, it has a very high resistance. Current can only flow through a standard diode in one direction.

Causal links and links by association

The link between the light intensity and resistance of an LDR is causal. This means the increase in light intensity causes a change in resistance. It is not quite so straightforward for the filament lamp. In this case, the link between potential difference and resistance is due to association: the increased potential difference causes an increase in temperature. It is the increase in temperature that causes the change in resistance.

3 You are designing the electrical system for a new yacht. Which of the components in photographs C to F would you need in a circuit to:
   a automatically switch the lights on when it gets dark
   b make a warning buzzer sound if the engine gets too hot
   c adjust the volume of the radio
   d stop you connecting the battery the wrong way round?

4 Some types of electronic equipment, such as computers, need a cooling fan. Explain why this is necessary, in as much detail as you can.

5 Draw up a table to describe the different components mentioned on these two pages. In your table, include a description of how the resistance of each component can be changed.

Using circuits

By the end of this topic you should be able to:

- apply the principals of basic electrical circuits to practical situations.

We use electrical circuits in many ways. For example, any boats sailing at night must show red, green and white navigation lights so that other boats can see where they are.

A Navigation lights on a hovercraft.

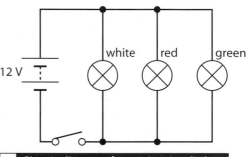

B Circuit diagram for navigation lights.

1 Look at circuit B.
  a Are the yacht's lights connected in series or in parallel?
  b What is the potential difference across the red light bulb when the lights are switched on?
  c What is the potential difference across the green bulb when the lights are switched on?
  d What would happen to the white light if the red one got broken?
  e Explain why the lights are connected this way in as much detail as you can.

Andy Hindley wants to leave his yacht anchored in a busy harbour for a few days. He wants to make a circuit that will automatically switch the navigation lights on when it gets dark. He builds circuit C to test his ideas.

Circuit C is called a **potential divider circuit**, because the potential difference supplied by the battery is divided between the two resistors. If both resistors have the same resistance, then the potential difference across each one will be 6 V.

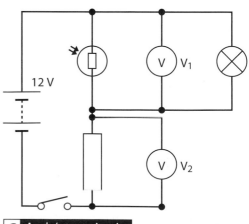

C Andy's test circuit.

The power of an appliance can also be calculated from the current it uses and the potential difference of the supply.

power (watts) = current (amps) × potential difference (volts)

Example

A light in a yacht needs a current of 2 A from a 24 V battery. What is its power?

power = current × potential difference
     = 2 A × 24 V
     = 48 W

This means that the light transfers 48 J of energy every second.

B

2 Calculate the power of these items:
  a an electric iron which uses a current of 4 A from the 230 V mains supply
  b a torch bulb which uses a current of 0.1 A from a 6 V battery
  c a hairdryer which uses a current of 3 A.

C The power rating of an appliance is marked on it.

We can use the equation for power to work out what size fuse a particular appliance should have. Fuses are normally available for the following currents: 3 A, 5 A, 13 A.

$$\text{current (amps)} = \frac{\text{power (watts)}}{\text{potential difference (volts)}}$$

Example

A hair straightener has a power rating of 40 W. What fuse should it have fitted?
Remember that mains appliances use a potential difference of 230 V.

$$\text{current} = \frac{\text{power}}{\text{potential difference}}$$
$$= \frac{40\text{ W}}{230\text{ V}}$$
$$= 0.17\text{ A}$$

The plug should be fitted with a 3 A fuse.

3 Work out what size fuse each of these items should have:
  a a lamp with a 100 W bulb
  b a 0.5 kW vacuum cleaner
  c an electric fire with two bars, where each bar has a power of 1 kW
  d a toaster with a power rating of 850 W.

4 **a** Explain what 'power' means and how it can be calculated from current and potential difference.
  b Explain how this equation can be used to work out the fuse needed for a particular appliance.

Investigative Skills Assessment 2

'Wires-R-us' supply electrical wire. Their testing laboratory has been asked to confirm that the resistance of the wire is proportional to length in a particular batch of wire.

The tests were carried out by Adam and Zoe. They both used the same apparatus, measured the current and potential difference for different lengths of wire and then calculated the resistance for each length.

The graph shows their results.

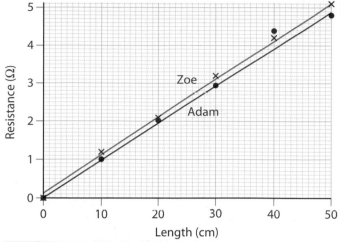

D Graph of Adam and Zoe's results.

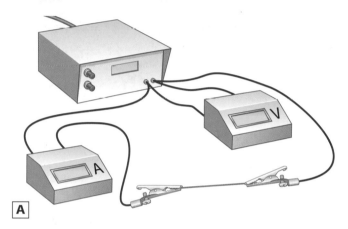

A

Length (cm)	Potential difference (V)	Current (A)	Resistance (Ω)
10	3	3.01	1.00
20	3	1.48	2.03
30	3	1.02	2.94
40	3	0.72	4.40
50	3	0.62	4.80

B Adam's results.

Length (cm)	Potential difference (V)	Current (A)	Resistance (Ω)
10	12	10.29	1.20
20	12	5.62	2.10
30	12	3.79	3.20
40	12	2.84	4.20
50	12	2.38	5.10

C Zoe's results.

1 Adam and Zoe were investigating the effect of the length of a piece of wire on its resistance.
  a Name **two** other variables they must keep the same to make sure their test is fair. (*2 marks*)
  b Describe how they could control one of these variables. (*1 mark*)
  c What was the interval between the values of their independent variable? (*1 mark*)

2 a One of Adam's results does not fit the pattern. Which result is this? (*1 mark*)
  b How could you find out if Adam's readings were wrong due to experimental error or if he made a mistake in his calculations? (*2 marks*)

3 a Is the resistance of the wire proportional to its length? Write 'yes' or 'no', and explain your answer. (*1 mark*)
  b Do you think there are enough results to make this a reliable conclusion? Write 'yes' or 'no', and explain your answer. (*1 mark*)

The two sets of results are not exactly the same. The difference could have been caused by Adam or Zoe not carrying out their experiment properly.

4 a Do the results show a random error or a systematic error? (*1 mark*)
  b Explain your answer to part **a**. (*1 mark*)

5 How can you tell that the error was not caused by a faulty measuring instrument? (*1 mark*)

6 a What was different in the way that Adam and Zoe carried out their investigation? (*1 mark*)
  b ✎ Explain how this difference could have given the different results. (*4 marks*)

Total = 17 marks

1 A boy is cycling through woods. He moves 60 metres at a constant speed, taking 20 seconds. He then stops for 10 seconds before moving at the same speed again for another 30 seconds.

   a Sketch a distance–time graph for this minute of the boy's ride. *(4 marks)*

   b If the times given above were measured by another boy watching, give a source of error in the timings. *(1 mark **HSW**)*

   c What other way could the results have been shown? *(1 mark **HSW**)*

2 **a** Annabeth wants to do a parachute jump but the pilot will only take people who weigh less than 1000 N. Her mass is 72 kg. Calculate Annabeth's weight. (g = 10 N/kg.) *(2 marks)*

   b When Annabeth jumps from the plane, she falls faster and faster until she reaches a constant velocity.

    i What happens to the force of air resistance as Annabeth accelerates? *(1 mark)*

    ii What can you say about the forces on Annabeth when she has reached the constant velocity? *(1 mark)*

3 Gareth did an experiment to show how seatbelts can reduce injury in a crash. He placed an egg on a model car and rolled it into a model wall.

   a Gareth did the experiment a second time, with the egg taped onto the car. The tape acted like a seatbelt.

   When the egg crashed this time, the tape stretched a little, holding the egg in place. Explain how this is safer for the egg. *(2 marks)*

   b Gareth repeated his experiment but replaced the wall with another, stationary, model car.

   The mass of the egg and the car together is 0.5 kg, and Gareth rolls it at 0.4 m/s.

    i Calculate the momentum of the egg and car at the start. *(2 marks)*

    ii State the total momentum after the collision. *(1 mark)*

   c What can you say about the kinetic energy of the egg and car if Gareth rolls them faster? *(1 mark)*

4 In a typical nuclear fission reaction, a uranium-235 nucleus absorbs a neutron and splits into two smaller nuclei.

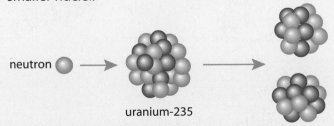

   a **i** Name **two** other things given off by this fission reaction. *(2 marks)*

    ii How could this start a chain reaction? *(1 mark)*

   b How is nuclear fusion different from nuclear fission? *(1 mark)*

5 The drawings show the inside of a plug and a cable.

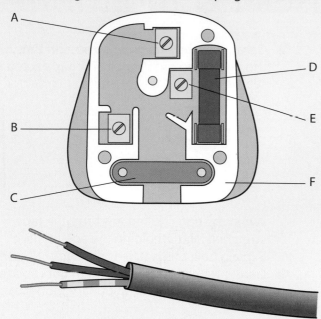

a Copy and complete this table to show how the wires should be connected to the plug. *(3 marks)*

Colour of wire	Name of wire	Which part should it be connected to?
blue		
green and yellow		
brown		

b Part F of the plug is made from plastic. Why is this material used? *(1 mark)*

6 Teresa is investigating static electricity. She has rubbed two rods with a cloth and hung them up. Some particles were transferred when the rods were rubbed.

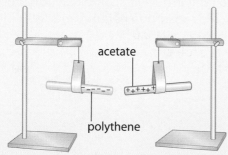

a Which particles were transferred? Choose from: ions, electrons, protons, neutrons. *(1 mark)*

b Were the particles transferred from the cloth to the acetate rod, or from the acetate rod to the cloth? Explain your answer. *(2 marks)*

c Look at the drawing of the two rods. Will they attract or repel each other? *(1 mark)*

7 Vijay sets up this circuit to measure the resistance of a resistor. He changes the potential difference by changing the setting on the power supply.

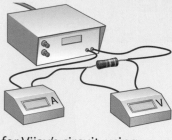

a Draw a circuit diagram for Vijay's circuit, using the correct symbols. *(2 marks)*

b To calculate the resistance, Vijay measures the current through the resistor and the potential difference across it. What piece of apparatus does he use to measure the current? *(1 mark HSW)*

c Vijay has several different instruments that he can use for measuring the current. They have different scales:

A measures from 0 to 1 amps
B measures from 0 to 5 amps
C measures from 0 to 10 amps

Explain how Vijay should decide which instrument to use. *(3 marks)*

d The table shows Vijay's results.

Potential difference (V)	1	2	3	4	5
Current (A)	0.08	0.11	0.25	0.34	0.41
Resistance (Ω)	12.50	18.18	12.00	11.76	12.20

i Which result is probably a mistake? *(1 mark HSW)*

ii Suggest how this mistake could have been made. *(1 mark HSW)*

iii What is the range of Vijay's results? (Ignore the incorrect result.) *(1 mark HSW)*

iv What is the mean value for the resistance? *(1 mark HSW)*

8 Tammy is investigating series and parallel circuits. She sets up this circuit.

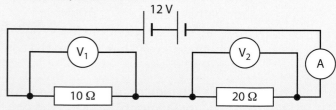

a Voltmeter V_1 reads 4 V. What will the reading on voltmeter V_2 be? *(1 mark)*

b Explain how you worked out your answer. *(1 mark)*

c What is the total resistance of the circuit? Explain your answer. *(2 marks)*

d Write down the equation that links power, current and potential difference. *(1 mark)*

e What is the power of Tammy's circuit? *(2 marks)*

Assessment exercises Higher

1 The graph below shows how the speed of a horse changed during the first 20 seconds of a race.

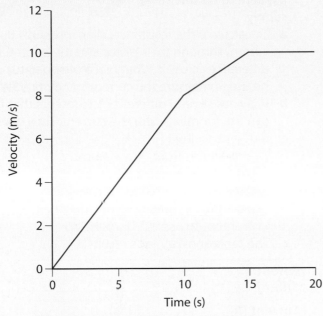

a How fast was the horse running after 10 seconds? *(1 mark)*

b What can you say about the horse's velocity between 15 and 20 seconds? *(1 mark)*

c What was the horse's acceleration over the first 10 seconds? *(2 marks)*

d How far did the horse run during the 20 seconds shown? *(2 marks)*

e If the speed was measured by timing the horse over a marked distance, suggest two possible sources of error in the results. *(2 marks **HSW**)*

f Why would it be important to carefully examine results like these if they were provided by the horse's owner? *(1 mark **HSW**)*

2 Gareth did an experiment to show how seatbelts can reduce injury in a crash. He placed an egg on a model car and rolled it into a model wall.

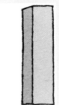

a Gareth did the experiment a second time, with the egg taped onto the car. The tape acted like a seatbelt. When the egg crashed this time, the tape stretched a little, holding the egg in place. Explain how this is safer for the egg. *(2 marks)*

b Gareth repeated his experiment but replaced the wall with another, stationary, model car which has a mass of 0.3 kg.

The mass of the egg and the car together is 0.5 kg, and Gareth rolls it at 0.4 m/s.

  i Calculate the momentum of the egg and car at the start. *(2 marks)*

  ii State the total momentum after the collision. *(1 mark)*

c The car collision takes 0.1 seconds and gives the second car 0.03 kg m/s of momentum. Calculate the force that acts between the cars. *(2 marks)*

d Calculate the kinetic energy of the egg and car before the collision. *(2 marks)*

3 Sketch a labelled diagram to show how a nuclear fission chain reaction might occur. *(2 marks)*

4 The drawing shows the inside of a plug.

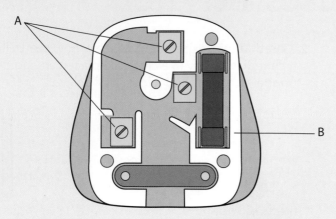

The parts labelled A are all made of metal, and part B is made from plastic.

a Why are the parts labelled A made from metal? Explain in as much detail as you can. *(2 marks)*

b Why is the part labelled B made from plastic? Explain in as much detail as you can. *(2 marks)*

5 Teresa is investigating static electricity. She has rubbed two rods with a cloth and hung them up. Some electrons were transferred when the rods were rubbed.

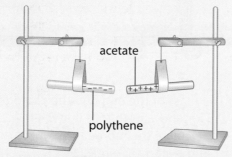

acetate

polythene

a Explain why electrons and not protons were transferred when the rods were rubbed. *(1 mark)*

b Were the particles transferred from the cloth to the acetate rod, or from the acetate rod to the cloth? Explain your answer. *(2 marks)*

c Look at the drawing of the two rods. Will they attract or repel each other? *(1 mark)*

d What would happen if Teresa had used two polythene rods? Explain your answer. *(2 marks)*

6 Vijay sets up this circuit to measure the resistance of a resistor. He changes the potential difference by changing the setting on the power supply.

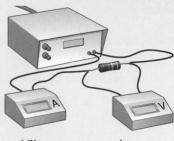

a To calculate the resistance, Vijay measures the current through the resistor and the potential difference across it. What piece of apparatus does he use to measure the current? *(1 mark HSW)*

b Vijay has several different instruments that he can use for measuring the current. They have different scales:

    A measures from 0 to 1 amps
    B measures from 0 to 5 amps
    C measures from 0 to 10 amps

Explain how Vijay should decide which instrument to use. *(2 marks HSW)*

c The table shows Vijay's results.

Potential difference (V)	1	2	3	4	5
Current (A)	0.08	0.11	0.25	0.34	0.41
Resistance (Ω)	12.50	18.18	12.00	11.76	12.20

  i Which result is probably a mistake? *(1 mark HSW)*

  ii Suggest how this mistake could have been made. *(1 mark HSW)*

  iii What is the range of Vijay's results? (Ignore the incorrect result.) *(1 mark HSW)*

  iv What is the mean value for the resistance? *(1 mark HSW)*

7 Tammy is investigating series and parallel circuits. She sets up this circuit.

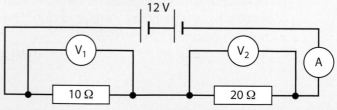

a What is the total resistance of the circuit? *(1 mark)*

b What will the readings on voltmeters V_1 and V_2 be? *(2 marks)*

c Explain how you worked out your answers to part **b**. *(2 marks)*

d Write down the equation that links power, current and potential difference. *(1 mark)*

e What is the power of Tammy's circuit? *(2 marks)*

Glossary

acceleration The rate of change of velocity; change in velocity in a certain time.

alpha (α) particle Radioactive particle which is two protons and two neutrons combined.

alternating current (a.c.) Current that changes direction many times each second.

ammeter An instrument for measuring the size of a current.

ampere (A) The unit for current, often shortened to amp.

atom Particle with no electric charge containing protons, neutrons and electrons. The smallest part of a chemical element.

atomic number Number of protons in the nucleus of an atom.

attract When two things try to move towards each other.

balanced Of forces that are equal and in opposite directions.

battery The scientific name for two or more cells connected together.

beta (β) particle Radioactive particle which is a fast-moving electron.

braking distance The distance, in metres, that a car's brakes need to stop it moving.

cable Used to connect appliances to the mains supply. Usually contains three insulated wires inside a plastic covering.

cell (electricity) The scientific name for a battery.

chain reaction A reaction in which the products cause the reaction to continue.

charge An amount of electricity.

circuit breaker A safety device which cuts off the current if the current flowing is too high.

collisions When moving objects hit each other, the total momentum is conserved.

components Parts of an electric circuit, such as bulbs, motors, cells, etc.

conductor A material that lets electricity flow through it easily.

conserved When a quantity is the same amount before and after an event.

consumer unit The place where the mains electricity comes into a house. A consumer unit includes fuses or circuit breakers.

coulomb (C) The unit for measuring charge.

current A movement of electrons (or other charged particles) through a material.

diode A component that lets electricity flow through it in only one direction.

direct current (d.c.) Electric current that flows in the same direction all the time.

discharge When something gets rid of a charge of static electricity.

distance–time graph A graph showing the position something has moved to at each moment in time.

earth wire A wire in mains cables used for safety.

elastic potential energy Also called strain energy; the energy stored in an object because of its shape.

electric circuit A complete loop of a conducting material with a cell and other components that electricity can flow around.

electron A particle inside atoms with a negative electric charge.

electrostatic charge An electric charge caused by an object gaining or losing electrons.

element Substance containing only one type of atom.

filament lamp A light bulb whose resistance increases when it gets hot.

fluid A liquid or gas.

force A push, pull or twist. An action that can change the movement of an object, or its shape.

frequency The number of times per second that something happens.

fuse (electricity) A piece of wire that melts and breaks a circuit if too much current flows through it.

gravitational field Area in which the force of gravity acts. On Earth, the strength of the gravitational field, g, is 10 N/kg.

gravity The attractive force produced by all objects that have mass.

hertz (Hz) The units for frequency. 1 Hz = 1 cycle per second.

in parallel Components connected so that the current splits up. Some of the current will flow through one component, and some will flow through the other.

in series A component connected in a circuit so that current must pass through it to the rest of the circuit.

insulating materials Materials that do not let electricity pass through them easily.

ion Electrically charged particle containing a different number of protons and electrons.

ionise Add or remove electrons from an atom to leave it charged.

isotopes Atoms of the same element with the same number of protons but a different number of neutrons (the same atomic number but a different mass number).

kinetic energy The energy an object has because it is moving.

light dependent resistor (LDR) A resistor whose resistance gets lower when light shines on it.

live wire The brown wire in a three-core cable. Its potential varies when an alternating current is flowing.

magnitude The amount or size of a quantity.

mass The amount of matter that makes up an object, measured in kilograms (kg).

mass number Number of protons plus the number of neutrons in an atom.

milliamp (mA) A unit for measuring small currents. 1 A = 1000 mA.

moderator A substance in a nuclear reactor which has the job of slowing neutrons down so that they can be absorbed by the nuclear fuel.

momentum (plural: momenta) A measure of the movement of an object: the product of its mass and velocity.

neutral wire The blue wire in a three-core cable. Its potential is usually zero compared to earth.

neutron A particle inside the nucleus of an atom with no electric charge.

newtons (N) The SI unit of force. (Note that the word should not be capitalised – the unit of force is correctly written in full with a small initial, n, but the abbreviation is a capital N.)

nuclear fission When a large, unstable atomic nucleus splits into smaller nuclei, releasing neutrons and much energy.

nuclear fusion When small nuclei collide and join together to form a larger nucleus, releasing much energy in the process.

nucleus (chemistry) The centre of an atom containing protons and neutrons.

ohm (Ω) The unit for measuring resistance.

oscilloscope Equipment used to show how voltage changes with time.

parallel circuit A circuit in which there is more than one path for the current to flow.

peak potential difference The highest potential difference in an alternating electricity supply.

H **period** The time for one complete cycle or wave.

photocopier A machine that uses static electricity to make copies of images or words in documents.

plum pudding model Thomson's idea of the structure of an atom as a lump of positively charged material with negative electrons scattered through it.

plutonium-239 A nuclear fuel that can be used for fission reactions in nuclear power stations.

potential difference The difference in the energy carried by the electrons before and after they have flowed through a component (sometimes referred to as voltage).

potential divider circuit A circuit that uses two resistors to change the potential difference supplied to other components.

power The amount of energy transferred per second.

proton A particle with a positive charge found in the nucleus of an atom.

reaction time The time taken for a driver to realise the need to stop the vehicle and start to press the brake pedal.

repel Push away.

resistance How easy or difficult it is for an electric current to flow through something.

resultant force The overall force when several forces are added together, accounting for their directions.

series circuit A circuit in which there is only one loop for the current to flow through, so the current goes through all the components, one after another.

smoke precipitator A device that uses static electricity to remove dust particles from the smoke passing up a chimney.

speed How fast something is moving; distance moved in a certain time.

static electricity Electric charges on insulating materials.

stopping distance The overall distance needed to stop a moving vehicle. The sum of the thinking distance and the braking distance.

terminal velocity The maximum velocity that an object falling under gravity, through a fluid, can travel at. It is reached when the drag forces on the object rise (with increasing speed) enough to balance the object's weight.

thermistor A component whose resistance gets smaller when it gets hotter.

thinking distance The distance travelled by a vehicle during the driver's reaction time.

three-pin plug A plug used to connect appliances to the mains electricity supply.

unbalanced Of forces, when the resultant force is not zero.

uranium-235 A nuclear fuel that can be used for fission reactions in nuclear power stations.

variable resistor A resistor whose resistance can be changed by turning a knob or moving a slider.

velocity Speed in a certain direction.

velocity–time (v–t) graph A graph showing the velocity of an object at each moment in time.

volt (V) The unit for measuring potential difference.

voltmeter A device for measuring the potential difference across one or more components.

watt (W) The unit for measuring power. 1 watt = 1 joule per second.

weight The force on an object caused by the gravity of the planet.

work The amount of energy transferred. The product of a force and the distance moved against the force.

Periodic Table

Key

relative atomic mass		
1		
H		
hydrogen		
1		

atomic symbol
name
atomic (proton) number

Groups: 1 2 3 4 5 6 7 0

Group 1	2											3	4	5	6	7	0
																	4 **He** helium 2
7 **Li** lithium 3	9 **Be** beryllium 4											11 **B** boron 5	12 **C** carbon 6	14 **N** nitrogen 7	16 **O** oxygen 8	19 **F** fluorine 9	20 **Ne** neon 10
23 **Na** sodium 11	24 **Mg** magnesium 12											27 **Al** aluminium 13	28 **Si** silicon 14	31 **P** phosphorus 15	32 **S** sulfur 16	35.5 **Cl** chlorine 17	40 **Ar** argon 18
39 **K** potassium 19	40 **Ca** calcium 20	45 **Sc** scandium 21	48 **Ti** titanium 22	51 **V** vanadium 23	52 **Cr** chromium 24	55 **Mn** manganese 25	56 **Fe** iron 26	59 **Co** cobalt 27	59 **Ni** nickel 28	63.5 **Cu** copper 29	65 **Zn** zinc 30	70 **Ga** gallium 31	73 **Ge** germanium 32	75 **As** arsenic 33	79 **Se** selenium 34	80 **Br** bromine 35	84 **Kr** krypton 36
85 **Rb** rubidium 37	88 **Sr** strontium 38	89 **Y** yttrium 39	91 **Zr** zirconium 40	93 **Nb** niobium 41	96 **Mo** molybdenum 42	[98] **Tc** technetium 43	101 **Ru** ruthenium 44	103 **Rh** rhodium 45	106 **Pd** palladium 46	108 **Ag** silver 47	112 **Cd** cadmium 48	115 **In** indium 49	119 **Sn** tin 50	122 **Sb** antimony 51	128 **Te** tellurium 52	127 **I** iodine 53	131 **Xe** xenon 54
133 **Cs** caesium 55	137 **Ba** barium 56	139 **La*** lanthanum 57	178 **Hf** hafnium 72	181 **Ta** tantalum 73	184 **W** tungsten 74	186 **Re** rhenium 75	190 **Os** osmium 76	192 **Ir** iridium 77	195 **Pt** platinum 78	197 **Au** gold 79	201 **Hg** mercury 80	204 **Tl** thallium 81	207 **Pb** lead 82	209 **Bi** bismuth 83	[209] **Po** polonium 84	[210] **At** astatine 85	[222] **Rn** radon 86
[223] **Fr** francium 87	[226] **Ra** radium 88	[227] **Ac*** actinium 89	[261] **Rf** rutherfordium 104	[262] **Db** dubnium 105	[266] **Sg** seaborgium 106	[264] **Bh** bohrium 107	[277] **Hs** hassium 108	[268] **Mt** meitnerium 109	[271] **Ds** darmstadtium 110	[272] **Rg** roentgenium 111							

Elements with atomic numbers 112–116 have been reported but not fully authenticated.

* The Lanthanides (atomic numbers 58–71) and the Actinides (atomic numbers 90–103) have been omitted.

Cu and **Cl** have not been rounded to the nearest whole number.

Data sheet

Reactivity series of metals

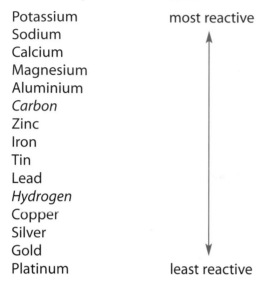

Potassium   most reactive
Sodium
Calcium
Magnesium
Aluminium
Carbon
Zinc
Iron
Tin
Lead
Hydrogen
Copper
Silver
Gold
Platinum   least reactive

(elements in italics, though non-metals, have been included for comparison)

Formulae of some common ions

Positive ions

Name	Formula
Hydrogen	H^+
Sodium	Na^+
Silver	Ag^+
Potassium	K^+
Lithium	Li^+
Ammonium	NH_4^+
Barium	Ba^{2+}
Calcium	Ca^{2+}
Copper(II)	Cu^{2+}
Magnesium	Mg^{2+}
Zinc	Zn^{2+}
Lead	Pb^{2+}
Iron(II)	Fe^{2+}
Iron(III)	Fe^{3+}
Aluminium	Al^{3+}

Negative ions

Name	Formula
Chloride	Cl^-
Bromide	Br^-
Fluoride	F^-
Iodide	I^-
Hydroxide	OH^-
Nitrate	NO_3^-
Oxide	O^{2-}
Sulfide	S^{2-}
Sulfate	SO_4^{2-}
Carbonate	CO_3^{2-}

Index